iRODS
User Group Meeting 2017 Proceedings

© 2017 All rights reserved. Each article remains the property of the authors.

9TH ANNUAL CONFERENCE SUMMARY

The iRODS User Group Meeting of 2017 gathered together iRODS users, Consortium members, and staff to discuss iRODS-enabled applications and discoveries, technologies developed around iRODS, and future development and sustainability of iRODS and the iRODS Consortium.

The three-day event was held from June 13th to 15th in Utrecht, Netherlands, hosted by Utrecht University and the iRODS Consortium, with over 90 people attending. Attendees and presenters represented over 50 academic, government, and commercial institutions.

Contents

Listing of Presentations 1

ARTICLES

iRODS workflows for the data management in the EUDAT pan-European infrastructure 5
Claudio Cacciari - CINECA
Robert Verkerk - SURFsara
Adil Hasan - SIGMA2
Javier Quinteros - German Research Centre for Geosciences (GFZ)
Julia Kaufhold - Max Planck Computing and Data Facility (MPCDF)

Neuroimaging Research Data Life-Cycle Management 13
Hurng-Chun Lee, Robert Oostenveld, Erik van den Boogert, Eric Maris - Donders Institute, Radboud University

Workflow-Oriented Cyberinfrastructure for Sensor Data Analytics 23
Arcot Rajasekar - University of North Carolina at Chapel Hill
John Orcutt, Frank Vernon - University of California, San Diego

Distributing the iRODS Catalog: A Way Forward 35
Terrell Russell, Michael Stealey, Jason Coposky, Ben Keller, Claris Castillo, Ray Idaszak - RENCI, UNC-Chapel Hill
Alex Feltus - Clemson University

FAIR Sequencing Data Repository based on iRODS 43
Felipe O. Gutierrez, Paul De Geest, Aldo Jongejan, Sjoerd Repping, J.T. van den Berg, Antoine H.C. van Kampen, Silvia D. Olabarriaga - Academic Medical Center of University of Amsterdam (AMC)
Diogo F.C. Patrão - A. C. Camargo Cancer Center, São Paulo

Swedish National Storage Infrastructure for Academic Research with iRODS 49
Ilari Korhonen, Dejan Vitlacil, Ilker Manap - KTH Royal Institute of Technology
Janos Nagy, Krishnaveni Chitrapu - Linköping University

A national approach for storage scale-out scenarios based on iRODS 55
Christine Staiger - SURFsara
Ton Smeele, Rob van Schip - Utrecht University

Davrods Enhancements as part of the Grassroots Infrastructure 65
Simon Tyrrell, Xingdong Bian, Robert P. Davey - The Earlham Institute

QueryArrow: Semantically Unified Query and Update of Heterogeneous Data Stores 71
Hao Xu, Ben Keller, Antoine de Torcy, Jason Coposky - University of North Carolina at Chapel Hill

Listing of Presentations

The following presentations were delivered at the meeting:

- **The iRODS Consortium in 2017**
  Jason Coposky, iRODS Consortium
  Terrell Russell, iRODS Consortium

- **iRODS Technology Update**
  Mike Conway, iRODS Consortium
  Terrell Russell, iRODS Consortium

- **The Apertif long term archive**
  Hanno Holties - ASTRON

- **Designing an institutional research data management infrastructure for the life sciences**
  Paul van Schayck - Maastricht UMC+

- **FAIR sequencing data repository based on iRODS**
  Felipe Gutierrez - AMC, Amsterdam

- **iRODS workflows for data management in the EUDAT pan-European infrastructure**
  Claudio Cacciari - CINECA

- **Neuroimaging research data life cycle management**
  Hurng-Chun Lee - Donders Institute

- **Provisioning flexible and highly available iRODS-based data service at the Euro-Mediterranean Center on Climate Change Foundation**
  Marco Mancini - CMCC

- **iRODS impact on science and data management**
  Ashok Krishnamurthy - RENCI

- **Using iRODS as a presentation layer for research data storage**
  Dan Hanlon - UCL

- **QueryArrow: Semantically unified query and update of heterogeneous data stores**
  Hao Xu - RENCI

- **I upgraded iRODS and I still have all my hair**
  John Constable - Wellcome Trust Sanger Institute

- **Lightning Talks: iRODS For Nix at UGA**
  Bruno Bzeznik, Oliver Henriot - Grenoble, GRICAD

- **Lightning Talks: iRODS HA Database Deployment**
  Adam Carrgilson - Norwich Bioscience Institute

- **Data Management and Analytics Solution**
  Navya Dabbiru - Tata Consultancy, Bayer Crop Science, Digital Farming

- **Using iRODS in Sugar deployments**
  Tony Anderson - Volunteer, OLPC

- **iRODS functionality within the Grassroots Infrastructure**
  Simon Tyrrell - Earlham Institute

- **Building a Dutch research data infrastructure**
  Frank Heere - SURFsara

- **A converged, fault tolerant, distributed parallel architecture for iRODS**
  Aaron Gardner - BioTeam

- **Building data systems with iRODS and Golang**
  John Jacquay - BioTeam

- **Swedish National Storage Infrastructure for academic research with iRODS**
  Ilari Korhonen - KTH
  Dejan Vitlacil - SNIC

- **A national approach for storage scale-out scenarios based on iRODS**
  Christine Staiger - SURFsara

- **Real-time data management in iRODS**
  Arcot Rajasekar - RENCI

- **iRODS user empowerment, a matter of "Sudo" microservices**
  Chris Smeele - Utrecht University

- **Making massive data sets nimble and flexible with IBM Cloud Object Storage**
  Anu Khera - IBM

- **Distributing the iRODS Catalog: A Way Forward**
  Terrell Russell - RENCI

- **Building a Trustworthy Repository**
  Reagan Moore - Retired (DICE)

- **Community Collaborations: DE, Syndicate**
  Nirav Merchant - CyVerse, University of Arizona

- **Metadata Templates**
  Mike Conway - RENCI

- **Perspective - Building a House**
  Dave Fellinger - iRODS Consortium

Articles

iRODS workflows for the data management in the EUDAT pan-European infrastructure

Claudio Cacciari
CINECA
Via Magnanelli 6/3,
Casalecchio di Reno (BO),
Italy
c.cacciari@cineca.it

Robert Verkerk
SURFsara
SURF Science Park Building,
Science Park 140
1098 XG, Amsterdam,
The Netherlands
robert.verkerk@surfsara.nl

Adil Hasan
SIGMA2
Abels gate 5, Trondheim,
Norway
adilhasan2@gmail.com

Javier Quinteros
German Research Centre for Geosciences
(GFZ)
Telegrafenberg
D-14473 Potsdam, Germany
javier@gfz-potsdam.de

Julia Kaufhold
Max Planck Computing and Data Facility
(MPCDF)
Gießenbachstraße 2
85748 Garching, Germany
julia.kaufhold@mpcdf.mpg.de

ABSTRACT

The European project EUDAT built a data e-infrastructure, called Collaborative Data Infrastructure (CDI), connecting 16 data and computing centres to support over 50 research communities spanning across many different scientific disciplines. One of the main challenges to implement such infrastructure was to enable the users to manage their data in the same way across the different data centres despite each centre has its own peculiarities at hardware, software and policy level. Therefore, EUDAT adopted iRODS to deal with this heterogeneity relying on its features:

- To define a common abstraction layer on top of the difference storage systems.
- To provide a shared set of software interfaces and clients to perform data management operations.
- To enforce a common set of policies.
- To federate different administrative regions.

On the other hand, each community has its own characteristics and often it requires specific customizations to cope with its data life cycle. Hence, beyond this common horizontal layer, through iRODS, EUDAT can offer the flexibility of a vertical integration with the community's tools and policies. In order to implement those policies and functions, in the context of this project, we extended iRODS with a set of rules and scripts, which form, together with the underlying software stack, the B2SAFE service. It allows the replication of data collections across different iRODS zones, takes care to assign a unique identifier to each data object and collection, to log every failed transfers and to store a minimal set of metadata together with the data themselves. The unique identifiers are stored in a de-centralized registry, called B2HANDLE, which makes them globally resolvable and persistent. In this article, we introduce the B2SAFE architecture and highlight the integration between iRODS and the B2HANDLE system and the corresponding workflows.

Keywords

Data Management, EUDAT, replication, B2SAFE, persistent identifier, handle

iRODS UGM 2017, June 13-15, 2017, Utrecht, Netherlands
[Authors retains copyright. EUDAT receives funding from the European Union's Horizon 2020 programme – DG CONNECT e-Infrastructure Contract No. 654065.]

INTRODUCTION

In the last years, the European Commission promoted and funded initiatives to strength the data and computing infrastructures, which supported the research communities. Before those initiatives, some infrastructures were already in place, others only planned, but the panorama was fragmented with lack of interoperability and the concrete risk of increasing maintenance costs. The EUDAT project [1] was born as an answer to those concerns [2], with the objective to build a real pan-European data e-infrastructure.

EUDAT adopted iRODS as one of the main component of its infrastructure, represented in Figure 1 and called Collaborative Data Infrastructure (CDI). The CDI, which is now fully operative, has an architecture based on services, which form an integrated suite, depicted in Figure 2. iRODS is part of the B2SAFE service [3], which supports the long-term data preservation.

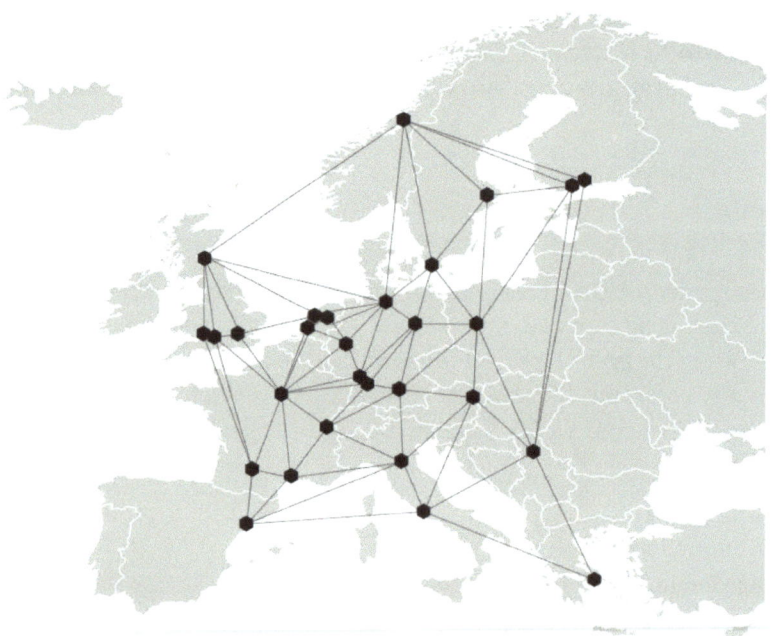

Figure 1. EUDAT Collaborative Data Infrastructure.

CHALLENGE

The Collaborative Data Infrastructure wants to offer a common experience to the users, providing them tools to manage their data in the same way across the different centers, which are part of the CDI. This is a great challenge, because it connects 16 data and computing centres to support over 50 research communities spanning across many different scientific disciplines and they are heterogeneous at hardware, software and policy level. The B2SAFE service relies on iRODS to overcome these obstacles, extending it through rule sets and scripts.

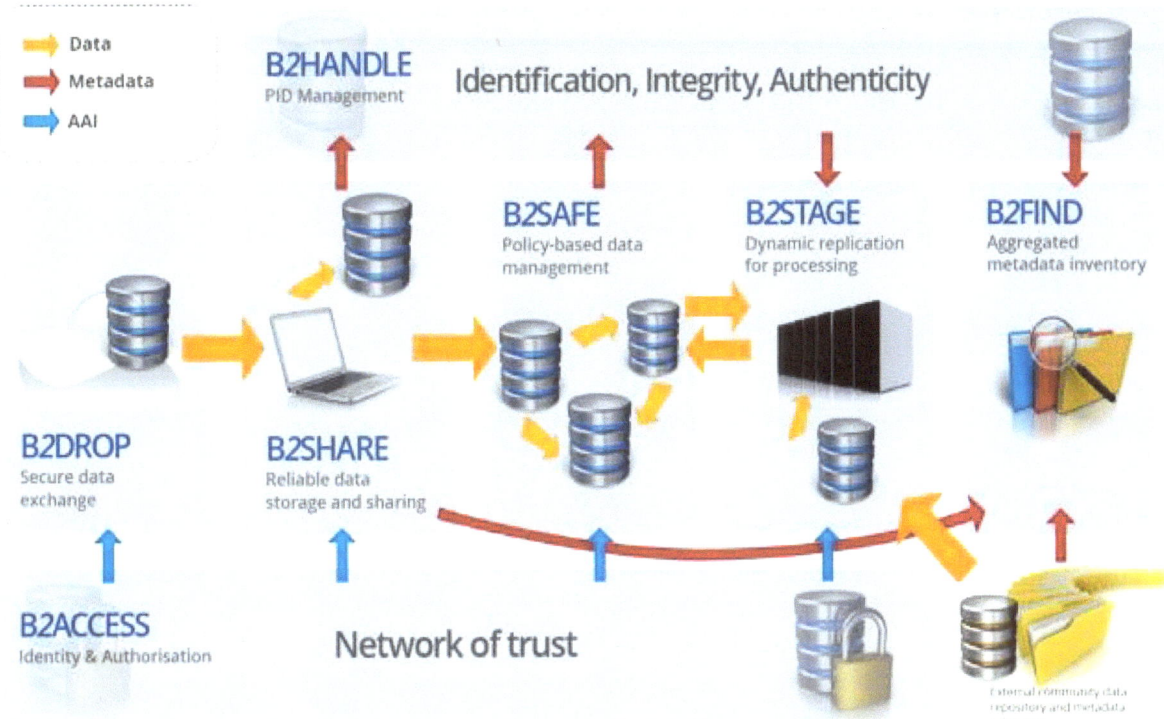

Figure 2. EUDAT data infrastructure service suite.

SOLUTION

The B2SAFE service provides a homogeneous and user-friendly interface to the user, dealing with four aspects related to critical differences among the various data centres:

1. Each CDI's data centre has different storage systems or, in the best case, different configurations of the same systems. The iRODS back-end abstraction layer allows hiding this complexity. In many cases, they are just plain file systems, but sometimes even tape libraries, or combinations of the two in a hierarchical solution. Part of this abstraction is also the common namespace, which simplifies both the data management for the users, and the back-end system maintenance for the administrators.
2. Because of this abstraction, it is possible to use a common set of software interfaces and clients to perform data management operations. In particular, B2SAFE allows the user to connect to it via iRODS icommands [4], WebDAV [5] and GridFTP. This last interface is an integration with the Globus GridFTP server [6], implemented through a specific library developed by EUDAT [7].
3. While the first and the second point address hardware and software heterogeneities, B2SAFE relies on the federation feature [8] to overcome administrative limitations. Each data centre in the CDI is an independent administrative unit; therefore, it overlaps perfectly with the concept of iRODS zone. Those zones are federated, allowing users to get access to multiple nodes of the infrastructures in a seamless way. However, the federation does not follow a full-mesh schema, where every zone is federated with each other, but it is configured as a set of islands, where each island, composed by two or more nodes, represents the storage area of a specific research community.
4. EUDAT defined a set of policies, implemented through iRODS rules, in order to enforce the data preservation best practices within the B2SAFE service. This set is packaged as an additional module [9], which is deployed on top of iRODS. In this way, the service offers common policies across the whole infrastructure.

IMPLEMENTATION

The B2SAFE service extensions to iRODS are implemented through rules and python scripts and can be grouped by functionality, as shown in Figure 3: logging, authorization, persistent identifiers (PIDs) [10] management, which rely also on software tools independent from iRODS; and data replication and error management, which are based completely on rules. Moreover, there is a set of utilities to support all the aforementioned groups.

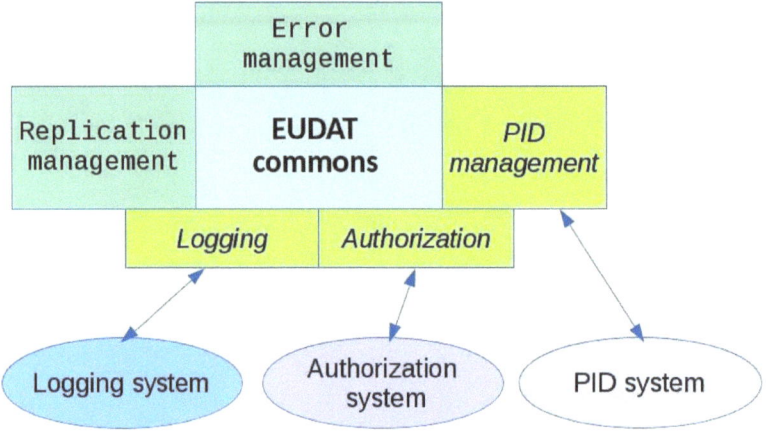

Figure 3. B2SAFE architecture.

Logging

The logging mechanism is independent from the core iRODS logging system and its purpose is to log the information related to the data replication rules. Currently the information is stored in files. In the future, the log messages could be queued into a messaging system.

Authorization

The authorization extension is a tool to control "who can execute which rule" and it is implemented through a rule (EUDATAuthZ(*user, *action, *target, *response)), which compares three attributes: the username, the "action" and the "target" against a set of assertions stored on a JSON (JavaScript Object Notation) formatted file and it returns true in case of match, false otherwise. For example:

```
acPreProcForExecCmd(*cmd, *args, *addr, *hint) {
    if (*cmd != "authZmanager.py") {
        EUDATAuthZ("$userNameClient#$rodsZoneClient", *cmd, *args, *response);
    }
}
```

The above code, placed in the rule set core.re, enforces the permission to execute external commands according to the assertions defined in the assertion file, which could be like the following one:

```
{
 "assertion 1":
        { "subject": [ "jack, james, master*" ],
          "action":  [ "getLDAPattributes.py" ],
          "target":  [ "password" ]
        }
```

}

Which means that the users "jack", "james" and all those beginning with word "master" are allowed to execute the command "getLDAPattributes.py" with the argument "password". Even in this case the mechanism can be further developed, in the future, to query an external authorization system.

Persistent identifier management

The persistent identifiers (PIDs) management consists of multiple rules and a python based client (epicclient2.py), which is able to connect to an instance of the EUDAT B2HANDLE service [11]. A PID is a unique identifier, based on the Handle scheme [12], which is composed by a prefix and a suffix, for example: 842/f5188714-f8b8-11e4-a506-fa163e62896a, where 842 is the prefix. The B2HANDLE service is a distributed service, which allows publishing PIDs and making them globally discoverable, relying on a software component called Handle System [12], supported by DONA [13]. The client exploits the B2HANDLE library [14] to create, modify and search across PIDs. By design, the handle scheme permits to extend arbitrarily the set of attributes associated to a PID, called PID record. EUDAT defined a PID record profile to formalize the EUDAT extended attributes. In the Table 1, we reported just the mandatory ones, which are the only relevant for the current explanation.

Type	Description	Example
URL	The http address of the object.	http://eudat.data.center/b2safe/api/object/CINECA01/home/collectionOne/myobject
EUDAT/CHECKSUM	The MD5 checksum of the object	f63fe6ae1540199f65d6dd3d7048c46b
EUDAT/CHECKSUM_TIMESTAMP	The timestamp in ISO UTC/ZULU time of the update operation of the checksum	2008-09-05T16:30Z
EUDAT/FIXED_CONTENT	Boolean value to show if the content of the object is immutable (true) or can be updated without changing the PID (false)	True
EUDAT/FIO	First Ingested Object: the PID pointing to the location of the first CDI node which has ingested the object	11100/785973e1-cd53-4c2f-bfe9-ed60d355725b
EUDAT/PARENT	The PID pointing to the parent element in a replication chain	11100/563409ba-22c1-4187-a162-5859560f721d
EUDAT/ROR	A pointer to the community's Repository of Record element in a	Community/locator/xyz

		replication chain. It can be a PID or any other identifier chosen by the community	
	EUDAT/REPLICA	A list of PIDs pointing to the replicas in a replication chain	11100/d4012a53-ca78-4345-a818-bc2a96408765, 99000/28c50c7c-65d3-436a-bd69-a333d82df192

Table 1. EUDAT PID record profile: mandatory attributes.

Replication

As mentioned before, B2SAFE's main objective is to enforce policies for the long-term data preservation. In this context one of the most important strategies to keep the data safe and support disaster recovery scenarios, is the replication of data to multiple sites, geographically distributed. Besides, the data replication is a way to optimize the data exploitation. Because many of the CDI's data centers offer computing resources, therefore, the data replication allows moving the data close to those resources; and many scientific communities are distributed across Europe, hence having the data close to their institutions improve their accessibility.

iRODS offers already replication mechanisms, but within the same zone. We needed to replicate data sets across different zones, which implies to deal with a certain number of issues related to the tracking of the replicas, the fault tolerance, the data integrity and the performance. Thus, we defined a rule called *EUDATReplication*, which relies on all the aforementioned extensions. The rule can be triggered client-side, with the "irule" command, but it is usually called within a policy enforcement point in "core.re", so that it is triggered when a new object or a new collection is uploaded to a specific path. The rule can receive as input the path either of an object or of a collection and replicate it to the proper destination. This operation relies on the PIDs management rules and scripts to assign a PID to the source, the destination and link them together using the attributes of the EUDAT PID record profile, as depicted in Figure 4.

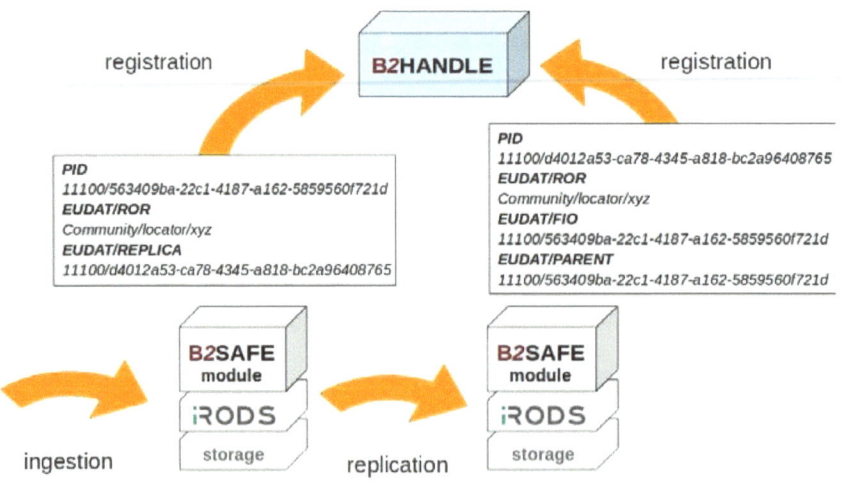

Figure 4. Replication process with PID attributes.

The replication sequence can involve multiple steps and supports different patterns. It could be a single chain of replicas and replicas of replicas, or, for example, have a star configuration, where each replica is copied directly from the master. Anyway, all the different patterns share a certain number of elements, which are tracked and form a double linked chain: each parent's PID record includes pointers to its replicas and each replica's PID record includes a pointer to the parent. Moreover, each replica's PID record includes the pointer to the first copy of the object ingested into the CDI (First Ingested Object, FIO) and, if it exists, the pointer to the master copy, stored outside the CDI, in the community's domain, known also as Repository of Records (RoR). This approach has two main benefits: it permits to the B2SAFE administrators to be always aware of the location and the number of copies of every object and collection stored on the infrastructure and it allows the users to find the data location that best fits their needs. In addition, in case of failure of one node of the CDI hosting a copy of the data, the user can always follow the pointers in the PID records to find another accessible copy.

However, this mechanism is meaningful only if the relation between the PID and the object's location is defined univocally (we are not considering here multiple versions of the same object for the sake of brevity): there must be only one PID for each object, therefore for each URL and CHECKSUM values in the PID record and vice versa. Because otherwise the meaning of persistent identifier as described in [10] is lost. The Handle server does not enforce this constraint intrinsically, therefore we have enforced it client-side in the B2SAFE PID management rules. The drawback of this further check is a decrement of the rate of the creation of new PIDs, during the PID registration. Rate that is also affected by security checks. Because, since the rules to create new PIDs are accessible to every user of a B2SAFE instance, we rely on the authorization extension to avoid that unauthorized users abuse of this feature. Usually, the performance of the whole replication procedure is not a critical aspect for the users, but sometimes, when the data set to be replicated scales to millions of objects, it is. In this case, the rule *EUDATReplication* is flexible enough to break the procedure into two sub-processes: the first one replicates the data, the second registers them. Thus the asynchronous PID registration can be scheduled later, once the data transfer is completed, allowing the B2SAFE administrators to find the most suitable period to perform it.

In the case of a failed replication, a certain number of checks are in place to identify the cause of the failure. Namely lack of permissions, failure in the PID registration and failure in the data transfer resulting in different size or checksum of the source compared to the replica. The replication rule saves the information about each failed task using the EUDAT logging extension. Then this information can be passed as input to another rule, *EUDATTransferUsingFailLog*, which is able to retry each failed replication. This asynchronous solution has been preferred to a synchronous one because often the reason of the failure lasts for a certain interval of time and it would be clearly useless to repeat the transfer within the same interval.

FUTURE WORK

The work described in the previous chapters can be improved along two directions. One is about the architecture: some of the components of the B2SAFE service are good candidates to be implemented as iRODS plugins. In particular, the authorization extension could benefit from this implementation in term of performance and better integration with the other rules. Other components could be, potentially, replaced by iRODS new features. It is the case of the logging mechanism, which could be superseded by the messaging framework. The other direction is related to the data management workflows. One of the PID record attributes is the object's checksum that is recorded to support the fixity feature, required to grant the data integrity as intended in [15]. The checksum of the master copy and of the replicas is compared periodically to verify that they are still coherent, however the B2SAFE administrator has to configure this procedure separately from the replication workflow. It is possible to achieve a better integration, abstracting some local system dependent details and automatizing part of the steps and creating in this way another building block to ease the definition of complex data policies.

CONCLUSIONS

Meeting the requirements of so many different scientific communities, it is really challenging for a pan-European data infrastructure like EUDAT. In this context, iRODS is important to support two types of integration. One aims to hide the complexity of the technology, providing a common layer on top of which is possible to build a shared set of policies for the data management. We call it horizontal integration, because it spans across data centers and communities forming a common base to integrate them at the same level, closing the gap, which often divides the first ones, more focused on the low-level technology layer, from the latter, more interested into the high level policies. Indeed, the heterogeneity is not only at technical level, but often also the maturity of the different communities differs in terms of awareness and knowledge of data management best practices and policies. To impose top-down solutions to the researches is usually counterproductive, hence it is necessary to help them customizing the services to make them closer to their daily practice and we call it vertical integration. The B2SAFE service is an example of this approach, because it implements some fundamental data management workflows, like the data replication and the assignment of globally discoverable identifiers, which can be used as building blocks from the users to define more complex and customized data policies.

REFERENCES

[1] EUDAT, https://www.eudat.eu, Visited last on 24.05.2017

[2] Lecarpentier, D., De Sanden, M.v., Wittenburg, P.: Towards A European Collaborative Data Infrastructure. Multi Science Publishing, doi:10.1260/2047-4970.1.0.233 (2012)

[3] B2SAFE service, https://www.eudat.eu/services/b2safe, Visited last on 24.05.2017

[4] iRODS icommands, https://docs.irods.org/4.2.0/icommands/user, Visited last on 25.05.2017

[5] WebDAV protocol, http://www.webdav.org, Visited last on 25.05.2017

[6] Globus GridFTP server, http://toolkit.globus.org/toolkit/docs/latest-stable/gridftp, Visited last on 25.05.2017

[7] B2STAGE DSI library, https://github.com/EUDAT-B2STAGE/B2STAGE-GridFTP, Visited last on 25.05.2017

[8] iRODS federation, https://docs.irods.org/4.2.0/system_overview/federation, Visited last on 25.05.2017

[9] B2SAFE module, https://github.com/EUDAT-B2SAFE/B2SAFE-core, Visited last on 25.05.2017

[10] Digital Preservation Handbook, second Edition, http://handbook.dpconline.org/, Digital Preservation Coalition (2015), Visited last on 25.05.2017

[11] B2HANDLE service, https://www.eudat.eu/services/b2handle, Visited last on 29.05.2017

[12] Handle system, https://www.handle.net, Visited last on 29.05.2017

[13] DONA, https://dona.net, Visited last on 29.05.2017

[14] B2HANDLE library, http://eudat-b2safe.github.io/B2HANDLE/index.html, Visited last on 29.05.2017

[15] Principles and Good Practice for Preserving Data, http://www.ihsn.org/sites/default/files/resources/IHSN-WP003.pdf, Interuniversity Consortium for Political and Social Research (ICPSR) (2009), Visited last on 29.05.2017

Neuroimaging Research Data Life-cycle Management

Hurng-Chun Lee	**Robert Oostenveld**	**Erik van den Boogert**
Donders Institute,	Donders Institute,	Donders Institute,
Radboud University	Radboud University	Radboud University
h.lee@donders.ru.nl	Karolinska Institute	e.vandenboogert@donders.ru.nl
	r.oostenveld@donders.ru.nl	

Eric Maris
Donders Institute,
Radboud University
e.maris@donders.ru.nl

ABSTRACT

Research Data Management (RDM) aims to improve the efficiency and transparency in the scientific process and to fullfil the requirements of the funding agencies and (local) regulations. Failures in reproducing some key empirical phenomena have resulted in the research process being questioned. In response to this, many research instititions have prioritized the development of RDM. In the neuroscience and neuroimaging domains, RDM is confronted with challenges in managing large volume and diverse data, which furthermore may contain sensitive personal information. At the Donders Institute (DI), we have developed an iRODS-based research data repository, an essential component for realising a RDM workflow that spans the whole research lifecycle. The objectives of this workflow are: (1) long-term data preservation for internal reuse, (2) documenting the analysis pipeline, allowing for collaboration, and reproduction of the published results, and (3) easy sharing of data and analysis pipelines with researchers around the world.

Keywords

Research data management, research life-cycle, neuroimaging data.

INTRODUCTION

Over the last ten years, researchers have experienced an increase in the pressure (by funding agencies, scholarly journals, and universities) to share their research data and to document the process via which they obtained their published results. This pressure asks for a Research Data Management (RDM) protocol and, given that almost all research data are stored digitally, a set of IT services that are needed to implement this protocol. In this paper, we describe such a RDM protocol and the associated IT services.

Research data are all the information that is (1) generated as a part of the research process and (2) on which a scientific report is/will be based. This definition of research data does not only include empirical data, but also simulated data, computer scripts for analysis and simulations, stimuli presented in experiments and the computer scripts for presenting them, etc. These data are generated in different stages of the research process, beginning by acquisition and ending by sharing. The RDM protocol that is described here pertains to the documentation of the data that are generated at *all* stages of the research process. Therefore, it is called *life-cycle RDM,* and it is to be compared with RDM restricted to the time of publication. Life-cycle RDM has the advantage that the act of documenting data becomes an integral part of the research process, with the sharing of documented data being a natural endpoint.

We implemented a particular protocol for life-cycle RDM in a large (600 researchers) and heterogeneous neuroscience institute, the Donders Institute (DI) for Brain, Cognition and Behavior. The DI acquires data of many different types, such as magnetic resonance imaging (MRI), different electrophysiological signals, whole-genome DNA, proteomics and transcriptomics data, limb movement trajectories, behavioural data, questionnaires, etc. These data are collected using a diverse set of instruments and machines, which often store the data in proprietary file formats. In addition, these data are analyzed using a diverse set of software tools, of which many also store their output in a proprietary file format.

In the following, we first describe our RDM protocol. This is followed by a description of the IT environment that is used to implement this protocol. We conclude this paper by giving an overview of the strengths and weakness of our RDM method, which is partially based on experiences with the method in a production environment.

THE RESEARCH DATA MANAGEMENT PROTOCOL

The objective of the Donders Institute (DI) Research Data Management (RDM) protocol is threefold: (1) data preservation for institute-internal reuse, (2) documenting the analysis pipeline, allowing for collaboration and reproduction of the published results, and (3) easy sharing of data and analysis pipelines with researchers around the world. To realise these objectives, three types of collections[1] have been defined: (1) Data Acquisition Collections (DACs), (2) Research Documentation Collections (RDCs), and (3) Data Sharing Collections (DSCs). These three types also correspond to three different phases of the research data life cycle: acquisition, analysis and reporting, and sharing. We have written a protocol that specifies how these collections are initiated, managed, built, closed and shared. Starting from this protocol, we have designed and built a digital infrastructure, the Donders Research Data Repository (DRDR) [1].

Collections are considered as resources that are provided to researchers upon request, just as lab space, access to MRI scanners, computing resources, research assistant hours, etc. These resources are managed at the level of a so-called Organisational Unit (OU). The DI has four OUs, that all manage their own resources. Collections are initiated by a member of the OU's administrative staff, the so-called Research Administrator (RA). Each collection belongs to a single OU, namely the OU of the RA that has initiated this collection. Upon initiation, disk quota is assigned to the collection, and the requesting researcher is granted the authorization to manage this collection (see further). Every collection includes a required attribute that specifies the research project with which this collection is associated. As a consequence, every OU must have an administrative system in place in which research projects are uniquely identified and in which resources are organised accordingly. Most OUs already had such a system in place, linking research projects to financial budgets.

Upon collection initiation, one or more researchers are assigned to this collection as so-called managers. From the perspective of the OU, these managers are responsible for building and curating the collection. This explicit responsibility implies that only members of that OU can be assigned as a manager. Within the DRDR, managers are authorized to assign other users to their collections. These other users can be assigned as manager, contributor or viewer. These roles map onto the familiar LINUX-authorizations own, write and read. A viewer can only read/download a collection's files, a contributor can also add/delete/modify these files, and a manager can also assign/remove other users to/from the collection (in a particular role). For a user to be assigned as manager to some of the OU's collections, the user has to be flagged as "eligible manager" in that OU[2]. As a contributor or viewer a user does not have to be specifically linked to one of the OUs. This feature accommodates on the one hand the formal responsibility (within the OU) and at the same time the frequent collaborations between researchers that belong to different OU's, or between researchers of which only a subset belongs to a OU that is represented in the DRDR.

---

[1] The term "collection" used in this paper refers to the DRDR collection. It should be distinguished from the iRODS collection. The DRDR collection is a conceptual container of folders and files. In practise, it is implemented as a iRODS collection in certain namespace hierarchy.

[2] The list of users eligleble to be manager within an OU is maintained by the RA of that OU.

The IT-system that implements the DRDR does not enforce how collections are to be built; this is rather specified in a written protocol. This written protocol specifies the collections primarily in functional terms (What should a collection viewer be able to do with it?), rather than in operational terms (Which files of which type must go in the collection, what must they contain, and how must they be organised?). Specifically, the protocol specifies that a DAC must contain all raw data (with "raw" meaning "without any manipulations that limit future analyses of these data") plus a description that would allow a well-informed colleague to make sense of the data. The written protocol is augmented by online documentation [2] in the form of a Frequently Asked Questions (FAQ) page, which contains concrete suggestions on how this functional requirement can be realised. Further, the protocol specifies that an RDC must document the scientific process, allow for the sharing of preliminary results within the project team (i.e. co-authors of a publication), and document the editorial and peer-review process. Again, this is augmented by a FAQ page that, in this case, contains concrete suggestions on how this documentation can be realised (e.g., by uploading analysis scripts). Finally, the protocol specifies that a DSC must contain all the relevant information (1) to reproduce the published results, and (2) to extend on these published results; this is again augmented by a FAQ page.

When a collection is complete, it can be closed, after which it cannot be changed anymore. Only a manager can close a collection. Closure of both a DAC and a RDC differs from closure of a DSC, this is crucial given that the DRDR can be used both for institute-internal research data management (realised by DACs and RDCs) and for sharing of data with the rest of the world (realised by DSCs). Specifically, only when closing a DSC, a persistent identifier is created via which this DSC can be accessed over the web. When closing a DAC or RDC, no persistent identifier is created.

Accessing data from a closed DSC cannot be done anonymously. To access a collection, users are required to be registered in the DRDR. To facilitate data sharing with a wide audience, the system allows users to register with the credentials of a social ID (e.g. LinkedIn, Google, Facebook, Twitter, Microsoft). As a DSC may only contain de-identified data, we obtained permission from our university's security officer to share data in this weakly-constrained fashion. To access DACs and RDCs (which may contain identifiable data, such as audio, video, MR scans, etc.) the authentication is more constrained, requiring evidence that the user is employed by an institute for scientific research (e.g., another university).

Accessing and re-using the data of a closed DSC is not unconditional. The conditions for access and re-use of the data are specified in a so-called Data Use Agreement (DUA). The minimum condition for access and re-use is that the data may not be used to try to identify the participants that have contributed the data. On top of that minimum condition, additional conditions may be imposed, such as conditions pertaining to the commercial and scientific use, and appropriate credit for acquiring and sharing these data. The DUA is selected by the DSC manager as one of the required collection attributes without which the DSC cannot be closed. Importantly, selecting an appropriate DUA, is the only way in which a DSC manager can control how the data may be reused; there is no individual review of data access requests by the manager. When a user (after registering with appropriate credentials, such as those of a social ID) accepts the DUA of a DSC, the DRDR automatically assigns this user as a viewer to the specific collection.

THE DONDERS RESEARCH DATA REPOSITORY

The objective of the Donders Research Data Repository (DRDR) is to provide a digital data management system that allows research data to be managed according to the protocol discussed above. Based on iRODS, the repository features:

- *A file-based system* that offers flexibility for managing unstructured data stored in various format.
- *A single and uniform namespace* that is understandable and familiar to researchers.
- *Access-controlled metadata management on collection level* providing essential information for documentation and searching.
- *User authentication* via trusted identity providers.
- *Role-based authorization* on collection and OU level.

- *Data replication* for disaster recovery.
- *Workflow automation and policy enforcement* to reduce administrative and operational effort.

Figure 1 shows the architecture of the DRDR system in three layers: storage system, data-management middleware and interfaces.

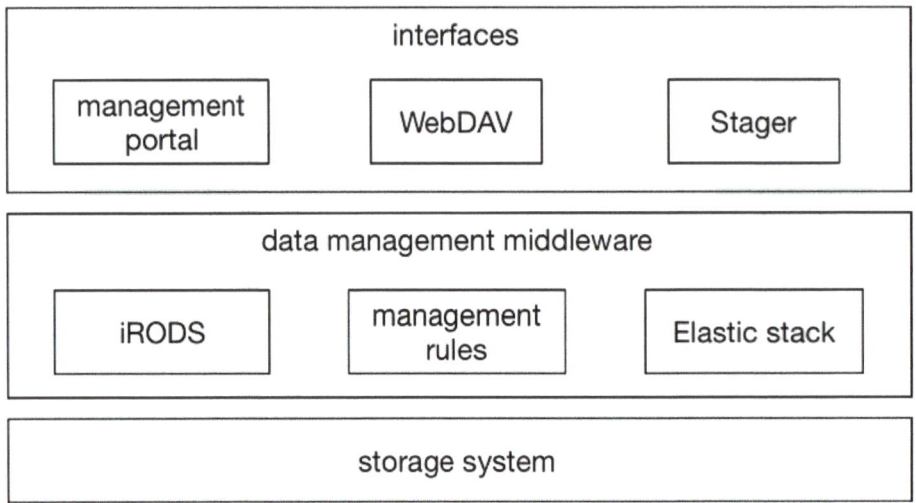

Figure 1: the DRDR architecture in three layers.

Storage system

The storage system is where the data are physically stored. It is mounted as a filesystem on the iRODS resource servers. In principle, the storage system can be of any type and specification. However, we anticipate some level of data duplication in different collections and thus rely on the de-duplication feature of the storage system to reduce the cost.

In the DRDR, two identical storage systems are installed at different locations for data safety and disaster recovery. For better control of dataflow and less dependency on storage functionality, data replication between the two storage systems is managed by the data management middleware.

Data management middleware

Leveraging on iRODS, the data management middleware implements the core functionality of the DRDR. We will discuss below the way iRODS is used in various aspects.

Resource arrangement

Using the tree metaphor and composable resources, the iRODS resources are arranged accordingly to maintain two data replicas in the system, and to reflect the administrative boundaries of different OU's. The arrangement is illustrated in Figure 2.

For the incoming data (i.e. the first replica), multiple coordinating (random) resources are created, each for an OU. Within each coordinating resource, storage (UNIX filesystem) resources are mapped to the filesystem mounting the first storage system. Quota may be configured on the filesystem to restrict the overall storage usage of an OU. With a proper design of the iRODS namespace (see later), the incoming data is controlled to flow accordingly to its OU-specific coordinating resource. Although we are using a single file server for now, this configuration allows OUs to grow with different storage (financial or technical) decisions independently without interference to other OU's and users.

As an "internal backup", resource for the second replica is made as a single coordinating resource consisting of a storage resource pointing to the second storage system. Data replication is performed by a delay rule triggered upon creation (or update) of the first replica.

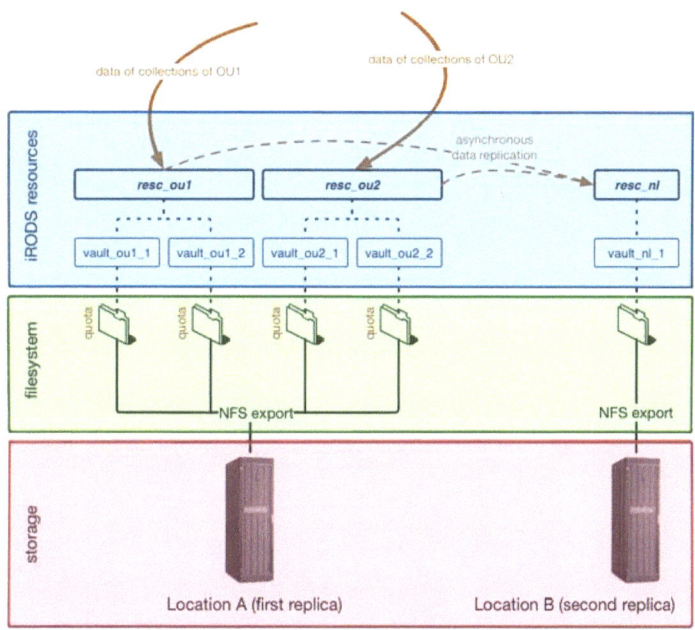

Figure 2: organisation of iRODS resources and dataflow in the DRDR system.

Collection namespace and role-based authorisation

The iRODS namespace of the DRDR collection is structured to reflect the hierarchy of organisation (O), organisational unit (OU) and the DRDR collections. Within a DRDR collection, the researcher has the freedom to organise data according to the specific (research) needs. Figure 3 shows an example namespace of a DAC belonging to the organisational unit DCCN.

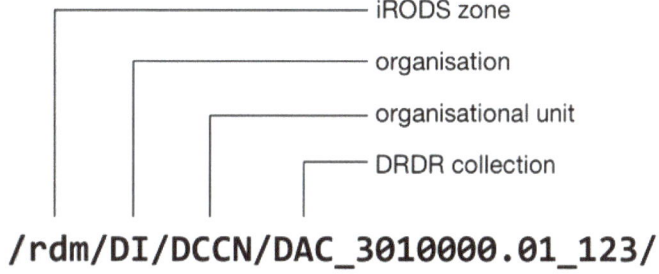

Figure 3: an example of the DRDR collection namespace.

The namespace hierarchy helps to implement the role-based authorisation defined by the protocol using the group-level access control. For instance, OU groups (e.g. 'ou_admin') are created to govern the access permission of the OU-level namespace; while three groups are made for each collection to control the permissions of 'manager', 'contributor' and 'viewer' roles. In this approach, granting/revoking access permission is simply achieved by adding/removing the user to/from a corresponding group.

Collection metadata

In the DRDR, metadata is only assigned to collections. On the collection-level namespace, metadata attributes are stored as the key-value-unit (KVU) triplets of iRODS. The attributes are largely derived from DataCite [3] with few control vocabularies specific to neuroscience and medical science (e.g. MeSH [4]). Attribute values containing data structure are represented as a JSON string.

Policy-enforcement points are adjusted to trigger data-management workflow when certain collection attribute is changed. For example, setting the 'state' attribute to 'closed' triggers a chain of actions to 1) set the collection to 'read-only', 2) clone the collection to a versioned snapshot, and 3) register the versioned snapshot with a global identifier (e.g. ePIC [5] or DOI) if the collection is a DSC.

Role-based authorisation is applied to metadata attributes according to the protocol. This is achieved by using the management rules (see later) given that iRODS doesn't support fine-grained authorisation on the KVU triplet.

Management rules for clients

A powerful feature of iRODS is the rule engine, which allows the data management policy and workflow to be customised. In the DRDR, we also use rules to create RPC-like client-server communications for high-level management such as editing collection attributes. The benefit of this approach is that complex logic and workflow managed on the iRODS server side are transparent to the web-application client. The management rules can be easily reused to build different management clients.

Figure 4 illustrates a client-server interaction using a management rule (i.e. rdmUpdateCollectionMetadata) for editing collection attributes. The client simply calls the rule with inputs; while on the server-side complex logics are wrapped together with the actual attribute update for, for instance, authorisation enforcement and verifying whether the values to be set are valid.

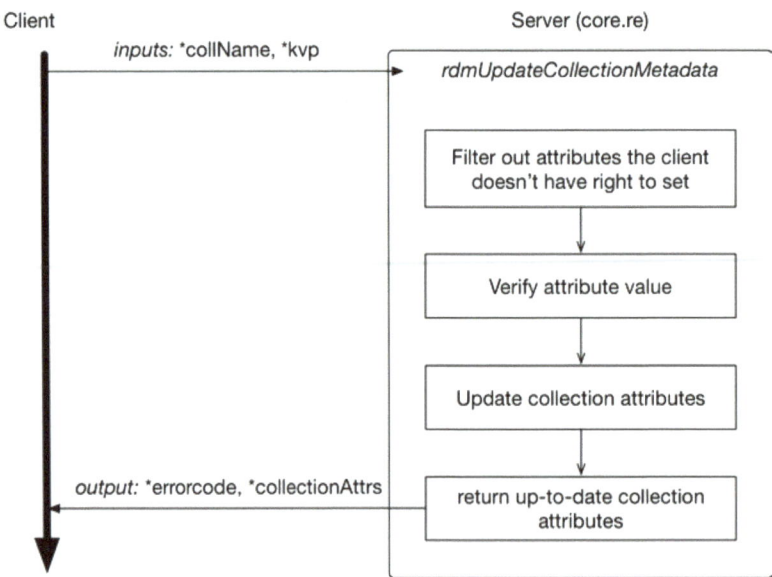

Figure 4: schematic illustration of the client-server interaction for updating collection attributes.

User provisioning and authentication

Users in the DRDR are provisioned as iRODS accounts when they sign-up the first time to the DRDR management portal via a trusted identity provider (IdP). Using the national IdP federation (i.e. SURFConext [6]), identities issued by Dutch research and educational institutions are supported. Social IDs are also supported with limited permission in DRDR. User attributes retrieved from IdP are stored in iRODS as user profile which is used to communicate with user (via e-mail attribute), and determine the user's eligibility in the system.

For security reason and to allowing auditing, users sign in to the management portal via a trusted IdP. This allows the system to record actions under an account that is traceable to an actual person. To access the data stored in a collection, the user has to retrieve the iRODS username and a short-term password from the management portal (after authentication). The iRODS username is used to interact with the WebDAV and iRODS data access interfaces. The short-term password prevents data from being accessible to a person whose authorisation has been revoked or expired by the trusted IdP.

Logging

User interactions with iRODS are logged as "events" written to the iRODS log file (i.e. rodsLog). The event content contains necessary details for auditing. Events in the log file are processed by Filebeat [6] and transferred into the Elastic stack [7]. Additional tools for accounting, notification and reporting are built on top of the Elastic stack. This dataflow is illustrated in Figure 5.

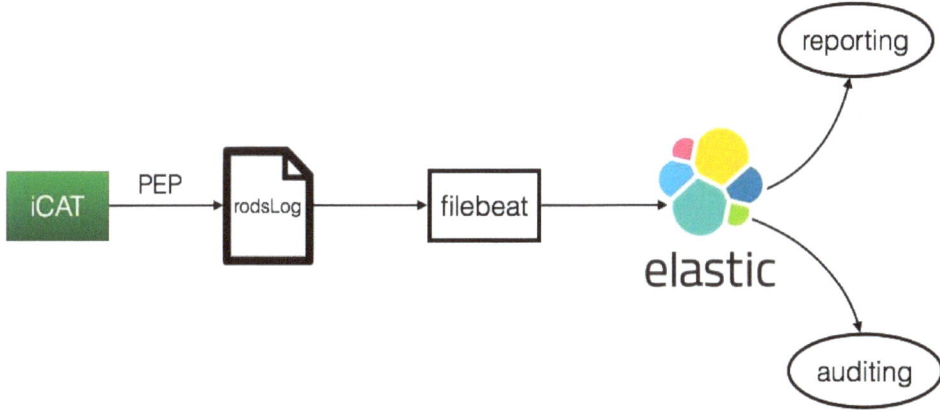

Figure 5: the dataflow of user events from iRODS to the Elastic stack.

Interfaces

One important feature in the DRDR interface is that it separates data access from the collection and user management. While the high-level rules consolidate the way management interfaces interact with iRODS, various data access mechanisms and interfaces are required for different scenario.

Management portal

The main management interface of the DRDR is a user-friendly web portal [1]. It is mainly used for managing user and collections attributes. Integrating with the federated identity provider, it is also used for user authentication and retrieving temporary credential for data access. Most of the functionalities in the manager portal require user authentication. The management portal also hosts the landing pages for published Data Sharing Collections which are publicly accessible. Under the hood, it makes use of the high-level rules to perform actions in iRODS.

WebDAV for easy data access

The main interface for transferring data in/out the DRDR is the WebDAV gateway using the Davrods [9] implementation. It allows researchers to use familiar tools to access data, and offers seamless integration with desktop by "mounting" collections as desktop drives.

The downside of WebDAV is that it is not trivial to maintain a reliable connection for transferring massive amount of data. Any failure along the connection will result in interruptions which then require human intervention. For researchers, these interruptions result in a considerable amount of effort, just to manage the transfers.

Stager for massive data ingestion and retrieval

The stager is a service implemented for data transfers between the DRDR and a local (high-performance) storage system, eliminating the need of researchers' effort in dealing with massive data transfers.

The stager provides a web interface for users to specify data transfer tasks in a graphical way similar to the file manager [10], and submit them to a task queue. Internally, an agent process performs the task of transferring data using the efficient "irsync" command and monitors the transfer progress. In case of failure, the task is rescheduled (up to a certain number of attempts). At completion, the user is notified by email about the result of the transfer.

The stager provides an efficient way for massive data ingestion and retrieval. Using the stager we implemented the automatic streaming of data from acquisition equipment (specifically the MRI and MEG scanners) to the DRDR. Upon the completion of a data-acquisition session in the lab, a data-transfer task is posted to the stager for ingesting the raw data to the DRDR. This automatic raw-data streaming into the DRDR provides not only a convenient automatic data stream for the researchers and an improvement in scientific integrity, but also an opportunity to organise the raw data into a standard structure (e.g. the BIDS standard [11]).

STRENGTHS AND WEAKNESSES

We outline two strengths and four weaknesses. The first important strength is that we have realised a protocol and associated IT environment that fits with the combined scientific-administrative workflow of a large and heterogeneous institute. Importantly, by treating storage on the DRDR as a resource for which researchers must apply just like other research facilities, we obtain adoption of the initial part of our RDM protocol. A second important strength is that DRDR provides the necessary functionality for (1) the sharing of the data of published papers, and (2) the implementation of the Data Management Plan (DMP) of a research grant. In a DMP, it must be described how the data will be handled, stored and shared in the different stages of the research project; most of these functions belong to the core functionality of the DRDR. Publishing papers and obtaining grants are important objectives for researchers, and this contributes to the user adoption.

A first weakness of DRDR is that it provides weak integration with the data analysis workflow using High Performance Computing (HPC) or using desktop computers. For instance, at this moment, we do not have a cross-platform solution allowing data analysis programs to access the DRDR collections as seamless as using a local storage. A second weakness of DRDR is that users can only be authorized for access to DACs and RDCs after they have authenticated via SURFConext as federated IdP service for research institutes and universities. Unfortunately, SURFConext operates on an opt-in basis: the DRDR has to be added by the IdP administrator for each of the research institutes or universities. This will only happen if employees of that organisation explicitly express to the local IdP administrator that they want to authenticate themselves in DRDR using their employer's IdP service. Only 9 of the approximately 100 institutions using SURFConext (which includes only 2 of the 14 Dutch research universities) has so far opted in. A third weakness of DRDR is that it does not implement standards for the collection content, simply because broadly applicable (cross-discipline) standards for collection content do not yet exist; the DI is an institute with a large variety of neuroscience disciplines. The absence of these standards hampers interoperability and reusability of the data. The fourth weakness is that the current DRDR interface doesn't provide enterprise-level searching-filtering-sorting functionality on collection metadata. This is due to limitations of the iRODS querying interface. For the moment, users can only filter collections on a few predefined attributes.

FUTURE DEVELOPMENT

Given the three collection types, DRDR provides a single repository that facilitates data management for both internal collaboration (via DAC and RDC) and data publication (via DSC).

In the perspective of internal collaboration, we are looking forwards developing an efficient integration between the DRDR system and computing facilities so that researchers can access collection content seamlessly for data analysis. Linked to this is an ongoing challenge of structuring the data in standard ways so that machine actions, such as automatic data processing, can be implemented within the collection.

For the data publication, our goal is to enable open data access following the FAIR principles [12]. Assigning persistent identifiers to published DSC's is the first accomplished step. In the future, we will address other aspects of the FAIR principle by, for example, exporting attributes of the published DSC's to FAIR data points with standard schemas (e.g. Dublin Core [13]). Given the sensitivity of the specific neuroimaging and medical data that we are managing, we note that not all FAIR aspects may be fully realised.

Furthermore, we will have to improve the system for better adoption and user friendliness. Speeding up data transfers, integration with a service interconnecting identity federations around the world (via eduGAIN [14]) and improving advanced searching-filtering-sorting functionality are important issues.

Although we believe that the method we have developed for managing neuroimaging data can be adopted by other research domains, we anticipate that some adjustments are needed to fit smoothly in a different domain. It is interesting to know how the current implementation will land in other research institutes. Feedbacks from the adoption process will help improve the system towards a more generic RDM solution.

CONCLUSION

We have structured the data management workflow in which both researcher and administration take part of responsibility. The workflow is specified by protocol and implemented by a digital data repository, and it pertains to the three data-management processes in the research lifecycle, namely the generation, documentation and publication of data.

The workflow defined in the protocol is content independent. It allows the protocol to be easily adopted by different research domains; while still offering flexibility for researchers to manage content towards their need.

The data repository is based on iRODS. The flexibility of iRODS allows us to leverage various iRODS features to implement the functional requirements derived from the protocol, and to provide feasibility for possible changes on the protocol.

Both the protocol and data repository are designed to provide a generic way of the research data lifecycle management. Thus, from our perspective, they can be either directly adopted by similar research institutions, or provided as a reference implementation for other research domain.

ACKNOWLEDGMENTS

We would like to thank the Management Board and the RDM steering group of the Radboud University for their guidance and financial support, the Radboud ICT Service Centre (ISC) for the technical support in developing the CMS portal of the repository, and the researchers at the Donders Institute for their invaluable input on design and usability aspects.

REFERENCES

[1] Donders Research Data Repository, **https://data.donders.ru.nl**

[2] The Frequently Asked Questions page of the Donders Research Data Repository, **http://www.ru.nl/donders/research/data/faq/**

[3] DataCite, **https://www.datacite.org**

[4] NCBI Medical Subject Headings (MeSH), **https://www.nlm.nih.gov/mesh/meshhome.html**

[5] ePIC, **http://www.pidconsortium.eu**

[6] Filebeat, **https://www.elastic.co/products/beats/filebeat**

[7] The Elastic stack, **https://www.elastic.co/products**

[8] SURFConext, **https://www.surf.nl/diensten-en-producten/surfconext/index.html**

[9] Smeele, T., Smeele C., Davrods, an Apache WebDAV interface to iRODS, Proceeding of iRODS User Group Meeting 2016, 41--48 (2016)

[10] Orthobox file managers, **https://en.wikipedia.org/wiki/File_manager#Orthodox_file_managers**

[11] Brain Imaging Data Strucutre (BIDS), **http://bids.neuroimaging.io**

[12] Wilkinson, M. D. et al., Comment: The FAIR Guiding Principles for scientific data management and stewardship, Scientific Data 3:160018 doi:10.1038/sdata.2016.18 (2016)

[13] The Dublin Core, **http://dublincore.org**

[14] eduGAIN, **https://www.geant.org/Services/Trust_identity_and_security/eduGAIN**

Workflow-Oriented Cyberinfrastructure for Sensor Data Analytics

Arcot Rajasekar
University of North Carolina
Chapel Hill, NC, USA
rajasekar@unc.edu

John Orcutt
University of California
San Diego, CA, USA
jorcutt@ucsd.edu

Frank Vernon
University of California
San Diego, CA, USA
flvernon@ucsd.edu

ABSTRACT

The iRODS middleware provides federation, virtualization, metadata integration and policy-oriented data management for static files. Real-time data streams (RTDS) from sensors and other sources pose a different challenge compared to static files. They are infinite in length and time line, comprise discrete packets of information which are time-specific, and the concept of byte-oriented i/o is not at all suited for accessing and managing real-time data streams. Moreover, because of the nature of the sources, the volume and velocity of the flow can be very low (temperature data) to very high (HDTV). They are also time-sensitive in two ways. The data must be captured immediately or they will be lost forever; and in many cases, the analysis has to be done immediately as the decision making can be time sensitive (eg. earthquake or tsunami detection). We have developed new features in the iRODS system to capture, store and archive RTDS. Our model captures RTDS into a continuum of discrete files and archives them in a few different standardized formats. It also provides packet-based and time-oriented access for replay of sensor data. By folding in the management of RTDS into iRODS we have extended the four main functionalities - federation, virtualization, metadata integration and policy-oriented automation – for real-time data management.

Keywords

Real-time data streams, iRODS, Datanet Federation Consortium, sensors.

INTRODUCTION

Sensors are electronic devices which detect or measure a physical property and convert them to electrical signals suitable for processing to be consumed by other electronic circuitry. Commonly detectable properties include biological, environmental, chemical, electrical, electromagnetic, mechanical, optical, radioactivity, etc. Sensors have been used for monitoring the environment [1,2,3,4], biological systems [5] and the human body [6]. A sensor grid [7,8,9] is a system that integrates sensor networks with grid computing and data grids. Sensor grids provide seamless access to distributed and heterogeneous sensors in a pervasive manner. It allows for applying large-scale computational power for analyzing sensor data, data fusion across multiple sensors, and developing novel algorithms for pattern recognition, sensor data discovery and decision making, using advanced techniques such as deep learning, machine learning, deep indexing, `data mining`, and distributed database processing. With the coming prominence of Internet of Things (IoT), more and more common everyday physical devices, buildings vehicles, and appliances are being embedded with electronic sensors leading to an enormous data volume that can be collected and processed [10,11,12]. Hence, there is a need for integrated data management of sensor data for collecting, storing, analyzing, sharing, discovery, and curating sensor data.

With the advent of IoT, Wearable Computers, Smart Cities and Connected Communities, and with large numbers of Science instruments being deployed, the amount of sensor-generated data is growing at a very fast pace. As of now, most of the sensor data are gathered, and analyzed in near-real-time, in situ or very close to the source and seldom archived for the long term (there are exceptions such as Incorporated Research Institutions for Seismology (IRIS),

Integrated Ocean Observing System (IOOS), NASA, etc.) This does not allow many options for reuse, replay and reanalysis of sensor streams. No sensor data management system is currently available, which can deal with millions of sensors, to store, access, analyze and manage sensor streams as is done with file storage and archives. The main reason is a lack of an ability to transport, store and retrieve sensor stream data at scale. When sensor streams are stored for later analysis, as in seismic data at the IRIS, or oceanographic data at IOOS, they are stored as files and access is to individual files through web links and ftp. Hence the problem is not only with storage, but also in retrieval and transport to users and sensor applications. The storage problem is compounded when dealing with millions of sensors.

Over the past 15 years there has been an explosion of sensor network data from many scientific disciplines including satellites in space observing magnetic fields and solar wind, meteorological networks for real-time forecasting and climate research, geophysical observations of earthquakes and tectonic motion, and physical oceanographic measurements of currents, temperature/salinity, waves and acoustic tomography/thermometry. With the advent of virtually ubiquitous networking, environmental sensor data are being streamed to a variety of locations for immediate application (e.g. tsunami detection). Today, seismic and environmental scientists continue to work with file-based systems including the extraction of data from the field and storage of sensor data. For example, field sensor data are first written to files, and at some later time the data are transferred to larger, community storage by copying the files and metadata over the Internet or actually mailing original physical media. The bulk of today's sensor data are managed through file-based systems, but streaming data analysis is quickly replacing the file-based approach even though the software continues to rely upon the traditional file approach. This fundamental mismatch needs to be addressed to meet the growing reliance on sensor stream data through research and development of a sensor-centric data system that provides end-to-end optimal performance.

The integrated Rule Oriented Data Systems (iRODS) [13,14,15,16] is a second-generation data grid that provides a collaboration environment for large-scale data oriented enterprises. iRODS promotes four concepts: virtualization, federation, automation and discovery (see Side Bar A). iRODS provides a rich client interface that supports a range of user-friendly interfaces for accessing data, ingesting and querying metadata, and for performing discipline-centric analysis and visualization through emerging social networking web applications. iRODS integrates and virtualizes distributed and heterogeneous data resources into a single logical file system (called the collection hierarchy) and provides a modular but uniform application processing interface to integrate new client-side applications as well as server-side data and compute resources.

> **A. iRODS Concepts**
>
> **Virtualization** allows users to create collections of dispersed data residing in distributed, heterogeneous resources and uniformly access them through single sign-on mechanisms. The data resources, the users and the access mechanism are represented by virtual name spaces that are mapped onto real objects. Virtualization implements technology independence and enables seamless access to data while hiding practical problems with authentication, authorization, arbitration and access to independently managed, heterogeneous resources. Virtualization also extends to compute services through containerization integrated with execution and data management. The iRODS system provides virtualization through mapping name spaces for users, resources, data, collections and micro-services (apps).
>
> **Policy-based Automation** enables customization and realization of complex resource management services at a fundamental level through computer actionable rules. Data resource managers, project leaders, and individual users can chain basic operations (micro-services) in order to define their own access pipeline, life-cycle management, sharing and disposition. The iRODS system used a rule engine to enforce policies stored in a rule base.
>
> **Federation** interconnects third party data, compute servers, and resources through the virtualization mechanisms. This enables a robust extensible framework for sharing resources owned and operated by third parties in a seamless manner. Multiple levels of interconnection are based on trust models and protocol brokering. iRODS system deployed tight and loosely coupled models as well as asynchronous federation mechanisms.
>
> **Metadata integration** is essential for discovery and sharing. Multiple types of metadata can be associated with data collections and need to be integrated through a common query mechanism. Metadata can range from key-value pairs, RDF and relational data to semi-structured and unstructured data. The iRODS system provided a common query mechanism through its logic-based QueryArrow system which provides a virtual query interface for both SQL and NoSQL databases.

The iRODS system has been used in multiple large-scale projects [17,18,19,20] and easily scales to 100s of millions of data objects in Petabyte storage systems and supports high-speed data transport including parallel streaming. In performing distributed data management, iRODS acts as a third-party intermediary providing authentication, authorization and auditing, and other functionalities that may or may not be supported by the underlying data resources. iRODS also provides optimized data movement protocols and rich support for metadata for data files as well as data collections. iRODS provides automation for data management as well as support for user-defined processing pipelines through its built-in distributed rule-engine. Administrators and collection owners can encode pipelines and policies as rules for managing and analyzing their data collections. The rule engine in iRODS provides a way to customize the community policies to meet the demands of each discipline and also encode trust relationships for sharing data across disciplines. Using the policy-based data management, one can encode data preservation and access functionalities such as data accession workflows, archival processes, dissemination processes, and analyses and access provisioning – all functions needed by large-scale digital sharing and curation systems. iRODS also provides a full range of services for long-term data management, including tracking replicas, versions, backup, and restoration. iRODS uses checksums to validate the integrity and supports automatic checking and repair of corrupted copies at user-defined intervals.

We have implemented a scalable sensor grid architecture that can be used to dynamically access packets of data in a stream from multiple sensors, and perform synthesis and analysis across a distributed network. Our system is based on the integrating iRODS with a new type of resource called the Antelope Real Time Data System (ARTS) [21], and providing virtualized access and handling to collections of data streams. The iRODS system brings to sensor processing features and facilities such as single sign-on, third party access control lists (ACLs), location transparency, logical resource naming, and server-side modeling capabilities while reducing the burden on sensor network operators. Rich integrated metadata support also makes it straightforward to discover data streams of interest and maintain data provenance. The workflow support in iRODS readily integrates sensor processing into any analytical pipeline. APIs for selecting, opening, reaping and closing sensor streams are provided, along with other helper functions to associate metadata and convert sensor packets into NetCDF and JSON formats. Near real-time sensor data including seismic sensors, environmental sensors, LIDAR and video streams are available through this interface [22]. We discuss the implementation of these features in some detail in the rest of the paper.

SENSOR DATA REQUIREMENTS

Sensor data have some peculiar properties compared to static data.. Unlike files, the sensors are highly distributed and their geographic location is an important property of their metadata and need to be captured. Another important feature of sensor data is that they are potentially infinite, but produced at discrete time intervals and referenced to a canonical time system. Hence the data stream from a sensor, unlike that from a file, can be unlimited and growing and cannot be defined in terms of bytes. Sensor streams comprise quanta of bytes called 'packets' and we can view a sensor data stream as a time-stamped series of packets. Hence, when accessing and storing data from sensors, one needs to deal with packets instead of byte buffers. These packets consist of a header, which can be used as an identifier (there are other metadata also part of the header) and a body (or payload) which contains the data. Moreover, a packet can be a complex, concentrated structure, having multiple sensor measurements either from many sensors collected over a period of time, or measurements taken by several co-located sensors at the same time and packed together in a single packet. Also, in many cases, the datum coming from the sensor can be a representation, example and electrical voltage value, which may be converted to the actual measurement of the physical entity before it can be stored or used in analysis. Hence, dealing with the ingestion and storage of each sensor stream may need to be customized and pre-processed.

Access to sensor data is normally done directly from sensors or at a concentrator (such as the ARTS systems). Currently most sensor processing is done in (almost) real time and seldom done afterward. When anyone wants to do post-facto analysis of archived sensor data, they are provided access to one or more files that contain the sensor data (stored in a standard format) and they have to deal with how to unpack that file and extract the time-series sensor data.

Replay and fast replay of sensor streams is very rare, but such a relay capability will be highly necessary if the interpreter wants to compare real time data with archived data.

If one wants to use data from multiple sensors in their analysis, unless they are available through the same concentrator, it is almost impossible to perform such analysis; this will need access to multiple sensor systems or concentrators (many of them have proprietary access control and authentication) as well as perform application-level time-alignment and data fusion. Hence providing capabilities for easy fusion of data from diverse and distributed sensor streams will be very useful for complex data analysis.

In our Datanet federation Consortium (DFC) [23,24,25] project, multiple usage models have been identified. The type of sensors that we need to access include marine, seismic, hydro and other environmental sensors, engineering sensors (as from smart buildings as well as infrastructures such as bridges), biological sensors, and diagnostic sensors such as MRI. The main needs of this group of scientists include ease of access to sensor data, export to standard formats so that it is easy to manipulate, access for archived sensor data, synchronized playback and integrated metadata for discovery and ease of integration into workflows and access through tools and applications. In order to meet these needs, we have developed an extension to the iRODS system enabling access, store, archive, discover, replay and analyze real-time data streams.

ANTELOPE REAL TIME SYSTEMS

Figure 1 Antelope Real Time System

The Antelope Real Time Systems (ARTS) [21] sensor data concentrator is used by multiple projects. ARTS (Figure 1) uses the concept of Object Ring Buffer (ORBs) to implement sensor data acquisition, transport, buffering, processing, archiving and distribution of environmental monitoring information. Antelope provides real time automated data processing and non-real time batch mode and interactive data processing. It has a built-in relational database for holding all raw data as well as processing results and other meta information. Antelope provides a comprehensive list of field interface modules for connecting with field sensor/digitizer/datalogger hardware to acquire data as well as state of health information and to control the field units. Antelope also has extensibility options in which application specific real time processing modules can be integrated easily for extracting information and knowledge from the raw data. Processing results are stored back into the same object oriented ring buffers as the raw data. These applications can be used for triggering new actions based on conditions/events emanating from one or more data streams. We use this facility to extract metadata automatically from a data stream. The Antelope System uses a relation-based database called Datascope which stores metadata in a relational schema as well as storing sensor data. It exposes a relational view of accessing sensor data as rows and allows time-interval querying. The ARTS system is used by multiple projects [26-29], including the SciON project [30], part of the DFC.

IRODS SENSOR INTEGRATION

The integration of the sensor data management in iRODS is done through an implementation of series of micro-services. The micro-services interact with the ARTS system and perform operations that provide access to sensor data from ARTS. Other micro-services were implemented to manipulate packet data, perform conversion and store them into the iRODS system. We discuss them below.

ARTS Micro-services

As mentioned earlier, ARTS stores and disseminates sensor packets in an object ring buffer. The ORBs store packets from multiple sensor streams. In order to access sensor packets, we need to connect to the ARTS systems at the host where it is running and perform stream-oriented operations. We have implemented micro-services for connecting to an ORB and disconnecting from it. The stream-oriented micro-services provide means to select a particular stream from the ORB and position the cursor for starting the read operation, using seek and position operations. The selection can also select one or more streams at the same time and can be used with wildcards for accessing data from multiple sensors, in interleaved time-series mode. Micro-services for getting the current packet, next packet in the time-series as well as installing a packet into the ORB are also implemented. Using these micro-services, one can open one or more streams and access packets in a row and then close the stream. Apart from these low-level micro-services for interacting with the ORB, two other micro-services called the *msiAntelopeGet* and the *msiAntelopePut* provide high level access for getting a stream in bulk mode and ingesting a stream back into the ORB systems. We also provide some heartbeat monitoring micro-services for checking the status of the ORB.

As mentioned before, a packet from a sensor stream can be quite complex with header and payload and the payload can also be quite complex. Micro-services are provided to decode and encode packets as well as unpack and repack a packet payload (using stuff and unstuff terminology which comes from ARTS). These micro-services along with another one which can perform format conversions of sensor data measurement form a suite of packet manipulation services that can be used in an iRODS workflow. Side Bar B provides brief explanations for these real-time sensor micro-services.

B. ARTS Micro-services

- Single Packet Microservices
  - msiAntelopeGet - get a packet
  - msiAntelopePut - put a packet
- Connection Microservices
  - msiOrbOpen - open an orb
  - msiOrbClose - close an orb
  - msiOrbTell - redirect to another orb
- Stream-level Microservices
  - msiOrbSelect - select streams
  - msiOrbReject - reject streams
  - msiOrbPosition - position read pointer by packetid
  - msiOrbSeek - position read pointer by skipping *n* packet
  - msiOrbAfter - position by time
  - convertExec - format conversion

- Packet Low-level Access Microservices
  - msiOrbGet   - get current packet
  - msiOrbReap   - get next packet
  - msiOrbReapTimeout - get next with timeout (return after timeout)
  - msiOrbPut - push a packet into stream
- Packet Manipulation Microservices
  - msiOrbUnstuffPkt - unpack a packet
  - msiOrbDecodePkt - decode a packet
  - msiOrbStuffPkt - pack a new packet
  - msiOrbENcodePkt - encode a packet
- Heartbeat Microservices
  - msiOrbStat – get info on streams
  - msiOrbPing – check on an ORB

Storage Formats

The sensor data stream is stored in the iRODS system as files. We provide three different formats for storing the time-series data. The first format is the compact format where the data packets are stored "as is". This format conforms to the one exported by the ARTS ORB system and can be easily ingested back into the ORB if needed for playback. The space needed for this is also quite low compared to other formats. A second format of storage is in the netCDF/HDF5 format called Common Data Language (CDL) format. This is human readable and is self-describing. We also provide storage in a third format using JSON. This format is useful for web-based apps. Figures 2 and 3 show some sample sensor data streams in the CDL and JSON formats. The schema for these two formats are defined such that the data can store a vector or a singleton measurement. Moreover, they are extensible as new data can be easily appended to the tail of these files.

```
netcdf barometric_pressure {
types:
 compound pressure_vector_t {
   double timestamp;
   float pressure ;
   float infrasound ;
 }; // barometric_vector_t
dimensions:
   time = UNLIMITED;
variables:
      pressure_vector_t barometric(time) ;
      barometric:standard_name = "two vector barometric pressure
data" ;
      barometric:long_name = "Barometric" ;
// global attributes:
   :srcname = "TA_003E/MGENC/EP1";
   :packettype = "waveform";
   :net = "TA";
   :sta = "003E";
   :chan = "LDO";
   :loc = "EP";
   :sampratepersec = " 1.000";
   :calib = "       1";
   :calper = "-1.000";
   :segtype = "5s";
   :nsamps = "120";
   :epochtime = "1446064294.9710000";
   :epochstarttime = "Wed 2015-301 Oct 28 20:31:34.97100";
   :epochendtime = "20:33:34.97100";
data:
barometric =
   {1446064294.9710000,  717022,  10159},
   {1446064295.9710000,  717021,   8821},
   {1446064296.9710000,  717023,  15918},
   {1446064297.9710000,  717026,  21402},
```

Figure 2 CDL Format: Pressure Data

```
{
 "packets":[
 {
   "srcname":"TA_J01E/MGENC/SM100",
   "pkttime":" 6/25/2015 (176) 0:30:23.968",
   "bytes":"535",
   "packettype":"waveform",
   "channels":[
   {
     "channum":" 0",
     "net":"TA",
     "sta":"J01E",
     "chan":"HNZ",
     "loc":"",
     "sampratepersec":"100.000",
     "calib":"       1",
     "calper":"-1.000",
     "segtype":"5s",
     "nsamps":"100",
     "epochtime":"1435192223.9683931",
     "epochstarttime":"Thu 2015-176 Jun 25
              0:30:23.96839",
     "epochendtime":" 0:30:24.96839",
     "data":[
        {"v":" -52727"},
        {"v":" -52729"},
        {"v":" -52729"},
        {"v":" -52731"},
```

Figure 3 JSON Format: Seismic VData

Demonstrating iRODS Sensor Workflows

The micro-services implemented for real-time data access and manipulation can be used along with other object-oriented micro-services in iRODS to perform many types of applications. Using the iRODS rule language we, can write applications that can be run interactively from the client or on the server side for continuous data reaping operations. We have developed several workflow programs to demonstrate the system capabilities. These include applications for (a) reaping *n* packets and storing them directly in the compact format in iRODS; (b) archiving one or more packets from a data stream in JSON format; (c) same as for CDL format; (d) ingesting packets into a sensor stream; (e) perform an orb2orb copy of a sensor data stream. (f) Access data from iRODS-stored files in CDL format through the iRODS Cloud Browser; (g) Show plots of data streams using the HDFViewer. Appendix A shows a few of the workflows that we have developed to showcase the application of these micro-services.

Demonstrating Real-time Sensor Data Access

One of the requirements for DFC is to show the streaming access of data from the iRODS system for sensor data. By its nature, the iRule command does not do streaming output and cannot deal with continuous data access. To perform this operation, we implemented a new iCommand called *isense* that can continuously reap packets from the iRODS-ARTS integration and show it on the screen.

    isense *orbHost   sensorName*

The command takes the ORB hostname, and a sensor name to continuously display the datapacket

    isense "anfexport.ucsd.edu:cascadia" "TA_J01E/MGENC/SM100"

The hostname maps to a particular ORB and the stream identified by the second parameter, in this case, is a seismic stream coming from the Anza Seismic Network. One can easily pipe this output to a stream processing application and perform real-time operations on the sensor stream.

We have used the isense command to reap sensor data and have piped it over the web using the websocketd [31] command-line tool to send a stream of data over to the web. At the web client side, we used the SmoothieChart [32] to continuously plot the sensor data stream on a web browser. This setup is shown in Figure 4.

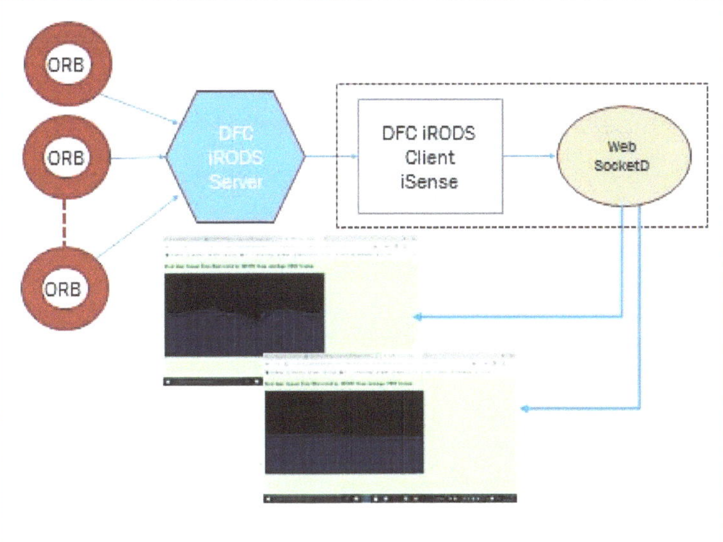

Figure 4 Web-access to Real-time Data Stream through iRODS

CONCLUSION

We have developed new features in the iRODS system to capture, store and archive RTDS. Our model captures RTDS into a continuum of discrete files and archives them in a few different standardized formats. It also provides packet-based and time-oriented access for replay of sensor data. By folding in the management of RTDS into iRODS we have extended the four main functionalities - federation, virtualization, metadata integration and policy-oriented data management – for real-time data. This extension to the iRODS system brings to sensor processing features and facilities such as single sign-on, third party access control lists (ACLs), location transparency, logical resource naming, and server-side modeling capabilities while reducing the burden on sensor network operators. Rich integrated metadata support also makes it straightforward to discover data streams of interest and maintain data provenance. The workflow support in iRODS readily integrates sensor processing into any analytical pipeline. APIs for selecting, opening, reaping and closing sensor streams are provided, along with other helper functions to associate metadata and convert sensor packets into NetCDF/HDF5 and JSON formats. With this extension, iRODS is well on its way to being a urban platform for smart cities and connected communities.

ACKNOWLEDGMENTS

This research is partially supported by NSF grant #0940841 "DataNet Federation Consortium".

REFERENCES

[1] Akyildiz, A, W. Su, Y. Sankarasubramaniam, E. Cayirci, "Wireless sensor networks: a survey," Computer Networks, Volume 38, Issue 4, 15 March 2002, pp. 393–422.

[2] Ho, C.K., A. Robinson, D. Miller and M. Davis, "Overview of Sensors and Needs for Environmental Monitoring," Sensors 2005, 5, pp 4-37.

[3] Ituen, I., G. Sohn, "The Environmental Applications of Wireless Sensor Networks," International Journal of Contents 2007, 3:4, pp 1-7, doi: 10.5392/IJoC.2007.3.4.001.

[4] Oliveira, L., J. Rodrrgues, "Wireless Sensor Networks: a Survey on Environmental Monitoring," JOURNAL OF COMMUNICATIONS, VOL. 6, NO. 2, APRIL 2011.

[5] Alemdar, H., C. Ersoy, "Wireless sensor networks for healthcare: A survey", Computer Networks, Volume 54, Issue 15, 28 October 2010, pp. 2688–2710.

[6] E. Egbogah, A. Fapojuwo, "A Survey of System Architecture Requirements for Health Care-Based Wireless Sensor Networks," Sensors 2011, 11, 4875-4898; doi:10.3390/s110504875.

[7] J. Yick, B. Mukherjee, D. Ghosal, "Wireless sensor network survey," Computer Networks, Volume 52, Issue 12, 22 August 2008, Pages 2292–2330.

[8] H.B. Lim, et al. Sensor Grid: Integration of Wireless Sensor Networks and the Grid, In Proc. of the IEEE Conference on Local Computer Networks 30th Anniversary (LCN'05), October 2005.

[9] M Richards; M Ghanem; M Osmond; Y Guo; J Hassard (2006), "Grid-based analysis of air pollution data", Ecological modelling, 194 (1-3).

[10] Xia, F., A. Yang, L. Wang, A. Vinel, "Internet of Things," Editorial, INTERNATIONAL JOURNAL OF COMMUNICATION SYSTEMS Int. J. Commun. Syst. 2012; 25:1101–1102.

[11] Atzori, L., A. Lera, G. Marabito, "The Internet of Things: A survey," Computer Networks, Volume 54, Issue 15, 28 October 2010, Pages 2787–2805.

[12] Bonomi, F., R. Milito, J. Zhu, S. Addepalli, "Fog computing and its role in the internet of things," Proceedings of the first edition of the MCC workshop on Mobile cloud computing, 2012, pp. 13-16.

[13] Moore, R., A. Rajasekar, "iRODS: Data Sharing Technology Integrating Communities of Practice", Proceedings of the 2010 IEEE International Geoscience and Remote Sensing Symposium, July 25, 2010.

[14] Moore, R.W., H. Xu, M. Conway, A. Rajasekar, J. Crabtree, H. Tibbo, "Trustworthy Policies for Distributed Repositories[2], Synthesis Lectures on Synthesis Lectures On Information Concepts, Retrieval, and Services, Vol 8 Issue 3, pp. 1-133.

[15] Moore, R., A. Rajasekar, H. Xu, "Extensible Generic Data Management Software", Journal of Open Research Software, July 2014.

[16] Rajasekar, A., M. Wan, R. Moore, W. Schroeder, S.-Y. Chen, L. Gilbert, C.-Y. Hou, R. Marciano, P. Tooby, A. de Torcy, B. Zhu, "iRODS – integrated Rule Oriented Data System", book in Synthesis Lectures on Information Concepts, Retrieval, and Services, Editor Gary Marchionini, Morgan Claypool Publishers, 2010.

[17] Cyverse, transforming science through data-driven discovery, **http://cyverse.org**.

[18] Hydroshare: Share and Collaborate, https://www.hydroshare.org.

[19] Chiang, G-T., P. Clapham, G. Qi, K. Sale, G. Coates, "Implementing a genomic data management system using iRODS in the Wellcome Trust Sanger Institute," BMC Bioinformatics, 2011, Volume 12, Number 1, Page 1.

[20] XSEDE, Extreme Science and Engineering Discovery Environment, `https://www.xsede.org/`.

[21] ARTS: Antelope Real Time Systems. **http://www.brtt.com/docs/ARTS.html**.

[22] Lindquist, K.G. and F.L. Vernon, D. Quinlan, J. Orcutt, A. Rajasekar, T.S. Hansen, S. Foley (2007). "The Data Acquisition Core of the ROADNet Real-Time Monitoring System". In Proceedings from Data Sharing and Interoperability on the World-wide Sensor Web (DSI 2007), Cambridge, MA (April 24, 48 pp.).

[23] DFC: Datanet Federation Consortium, **www.datafed.org**

[24] Billah, M., J. Goodall, U. Narayan, B. Essawy, V. Lakshmi, A. Rajasekar, R. Moore, "Using a data grid to automate data preparation pipelines required for regional-scale hydrologic modeling", Environmental Modeling and Software 78:31-39, March 2016.

[25] Moore, R., A. Rajasekar, "Reproducible Research within the DataNet Federation Consortium", International Environmental Modeling and Software Society 7th International Congress on Environmental Modeling and Software, San Diego, California, June 2014, http://www.iemss.org/society/index.php/iemss-2014-proceedings.

[26] USArray/Earthscope, **http://www.earthscope.org/**

[27] Anza Seismic Network, **http://eqinfo.ucsd.edu/deployments/anza.html**.

[28] HPWREN: High Performance Wireless Research and Education Network, **https://hpwren.ucsd.edu/**

[29] NEES: National Earthquake Engineering Simulation. **http://www.nees.org**

[30] SciON: Scientific Observatory Network, **http://scion-network.ucsd.edu/**

[31] WebSockets: the UNIX way, **http://websocketd.com/**

[32] Smoothie Charts: A JavaScript Charting Library for Streaming Data, **http://smoothiecharts.org/**

APPENDIX A: SINGLE PACKET REAP

```
reapAndConvertAntelopPacket {
#Get Packet
        msiOrbOpen(*orbHost,*orbParam, *orbId);
        msiOrbSelect(*orbId, *Sensor,*sresOut);
        msiOrbReap(*orbId, *pktId, *srcName, *oTime, *pktOut, *nBytes, *resOut);
        msiOrbDecodePkt(*orbId, *modeIn, *srcName, *oTime, *pktOut, *nBytes,
                  *decodeBufInOut);
        msiOrbClose(*orbId);
#Store Packet
        *SColl = *Coll ++ "/" ++ *Sensor
        *SFile = *SColl ++ "/" ++ "waveform.data";
        msiCollCreate(*SColl,"1",*STAT_1);
        openForAppendOrCreate(*SFile, *Resc, *D_FD);
        msiDataObjWrite(*D_FD, *decodeBufInOut, *WR_LN);
        msiDataObjClose(*D_FD,*STAT_2);
}

openForAppendOrCreate(*SFile, *Resc, *D_FD) {
# Sub Rule for appending if file already exists or creating header otherwise
        *SObj = "objPath=" ++ *SFile ++ "++++openFlags=O_RDWR";
        msiDataObjOpen(*SObj, *D_FD);
        msiDataObjLseek(*D_FD, *Offset,*Loc,*Status1);
}
openForAppendOrCreate(*SFile, *Resc, *D_FD) {
        msiDataObjCreate(*SFile, *Resc, *D_FD);
}
INPUT Coll="/rajaanf/home/rods/newsenstest",
        *Resc="destRescName=anfdemoResc++++forceFlag=", *Sensor= "TA_J01E/MGENC/SM100",
        *orbHost="anfexport.ucsd.edu:cascadia", *orbParam="", *modeIn=2, *Offset="0",
        *Loc="SEEK_END"
OUTPUT *pktId, *srcName, *oTime, *nBytes, *pktOut, *decodeBufInOut, ruleExecOut
```

APPENDIX B: ORB2ORB

```
orb2OrbReapedPacketIngestion {
#get a MGENC packet  from cascadia ORB and put it in demo  ORB
# also write also in a file to compare

# get the packet and the write into file
     msiAntelopeGet(*pktSelectInfo, *firstPktId, *lastPktId, NumOfPkts,*outBufParam);
    *SColl = *Coll ++ "/" ++ *Sensor
    *SFile = *SColl ++ "/" ++ "*firstPktId" ++ "_" ++ "*lastPktId" ++ ".data";
    msiCollCreate(*SColl,"1",*STAT_1);
    msiDataObjCreate(*SFile, *Resc, *D_FD);
    msiDataObjWrite(*D_FD, *outBufParam, *WR_LN);
    msiDataObjClose(*D_FD,*STAT_2);

# write to orb
    msiAntelopePut(*orbName, *srcName, *timeStamp, *outBufParam);
}
INPUT *pktSelectInfo="<ORBHOST>anfexport.ucsd.edu:cascadia</ORBHOST>
      <ORBSELECT>TA_J01E/MGENC/SM1</ORBSELECT><ORBWHICH>ORBOLDEST</ORBWHICH>
      <ORBNUMOFPKTS>1</ORBNUMOFPKTS><ORBNUMBULKREADS>1</ORBNUMBULKREADS>
      <ORBPRESENTATION>ONEPKT</ORBPRESENTATION>",
      *Resc="destRescName=anfdemoResc++++forceFlag=",
      *Coll="/rajaanf/home/rods/SensorData", *Sensor="TA_J01E_MGENC_SM1",
      *orbName="anfdevl.ucsd.edu:demo", *srcName="DFC_UNC/MGENC/T1", *timeStamp=""
OUTPUT *outBufParam, *firstPktId, *lastPktId, *NumOfPkts,  ruleExecOut
```

APPENDIX C: CONTINUOUS REAP

```
continuousReap {
  delay("<PLUSET>30s</PLUSET><EF>10m</EF>") {
      msiAddKeyVal(*KVP,"selectCriteria",*pktSelectInfo);
      msiAntelopeGet(*pktSelectInfo, *firstPktId, *lastPktId,
          *NumOfPkts,*outBufParam);
      *SColl = *Coll ++ "/" ++ *Sensor
      *SFile = *SColl ++ "/" ++ "*firstPktId" ++ "_" ++ "*lastPktId" ++ ".data";
      msiCollCreate(*SColl,"1",*STAT_1);
      msiDataObjCreate(*SFile, *Resc, *D_FD);
      msiDataObjWrite(*D_FD, *outBufParam, *WR_LN);
      msiDataObjClose(*D_FD,*STAT_2);
      msiAddKeyVal(*KVP,"firstPktId","*firstPktId");
      msiAddKeyVal(*KVP,"lastPktId","*lastPktId");
      msiAddKeyVal(*KVP,"numOfPkts","*NumOfPkts");
      msiAssociateKeyValuePairsToObj(*KVP, *SFile, "-d");
    }
    writeLine("stdout", "Delayed Rule Launched");
}
INPUT *pktSelectInfo="<ORBHOST>anfexport.ucsd.edu:cascadia</ORBHOST>
      <ORBSELECT>TA_M04C/MGENC/EP40</ORBSELECT><ORBWHICH>ORBOLDEST</ORBWHICH>
      <ORBNUMOFPKTS>8</ORBNUMOFPKTS><ORBNUMBULKREADS>4</ORBNUMBULKREADS>",
      *Resc="destRescName=anfdemoResc++++forceFlag=",
      *Coll="/rajaanf/home/rods/SensorData",
      *Sensor= "TA/M04C/MGENC/EP40"
OUTPUT ruleExecOut
```

Distributing the iRODS Catalog: A Way Forward

Terrell Russell
Renaissance Computing
Institute (RENCI)
UNC Chapel Hill
unc@terrellrussell.com

Michael Stealey
Renaissance Computing
Institute (RENCI)
UNC Chapel Hill
stealey@renci.org

Jason Coposky
Renaissance Computing
Institute (RENCI)
UNC Chapel Hill
jasonc@renci.org

Ben Keller
Renaissance Computing
Institute (RENCI)
UNC Chapel Hill
kellerb@renci.org

Claris Castillo
Renaissance Computing
Institute (RENCI)
UNC Chapel Hill
claris@renci.org

Ray Idaszak
Renaissance Computing
Institute (RENCI)
UNC Chapel Hill
rayi@renci.org

Alex Feltus
Clemson University
Clemson, SC
ffeltus@clemson.edu

ABSTRACT

In the last few years, researchers in academia and in both governmental and corporate sectors have become more interested in spanning greater physical distances within a single logical namespace for their files (iRODS Zone). However, connecting to a distant iRODS Catalog Provider presents a significant, if not unbearable, hurdle. This paper explores a solution to this new use case with a clustered database technology providing the iRODS metadata catalog (iCAT).

Keywords

iRODS, database, SQL, MariaDB, Galera, cluster, metadata

INTRODUCTION

An iRODS[1][2][3] use case was presented in which geographically disparate participants wanted to belong to the same iRODS Zone for ease of search and discovery, but they also wanted all iRODS servers to be provider nodes serving their own catalog. There was a desire to be able to decentralize the traditionally singular iCAT catalog database in a way that all participants could make use of whichever iCAT provider was closest to them rather than having to federate with separate iRODS Zones in distant physical locations. A multi-master solution would provide local authentication and improved metadata read performance while continuing to provide locality of reference for data at rest.

The initial requirements for such a system:

- Every iRODS provider node would contain the iCAT catalog and local storage space that could be uniquely assigned to that node as a resource.

- Files would be transferred to the iRODS provider node deemed closest to the point of file origination with respect to network latency and disk I/O metrics.

- All nodes must pass some sort of quality assurance test beyond the standard iRODS test suite which only exercises a single node via the default `demoResc` resource.

PROOF OF CONCEPT

The proof of concept solution being presented here uses MariaDB[4] configured as a Galera cluster to decentralize the iCAT catalog database across all participating iRODS provider nodes.

WAN replication will use ample latency values commensurate for an international WAN deployment.

A proof of concept testbed comprised of three iRODS provider nodes was deployed to form a single zone named `tempZone` within a MariaDB Galera cluster. Each node within the testbed is a single CentOS 7 VM, and can be configured to use differing latency values via NetEm[5] to simulate the kind of network traffic that would be experienced in a WAN configuration. Each VM is configured with a user account named **galera** which has rights to run Docker.

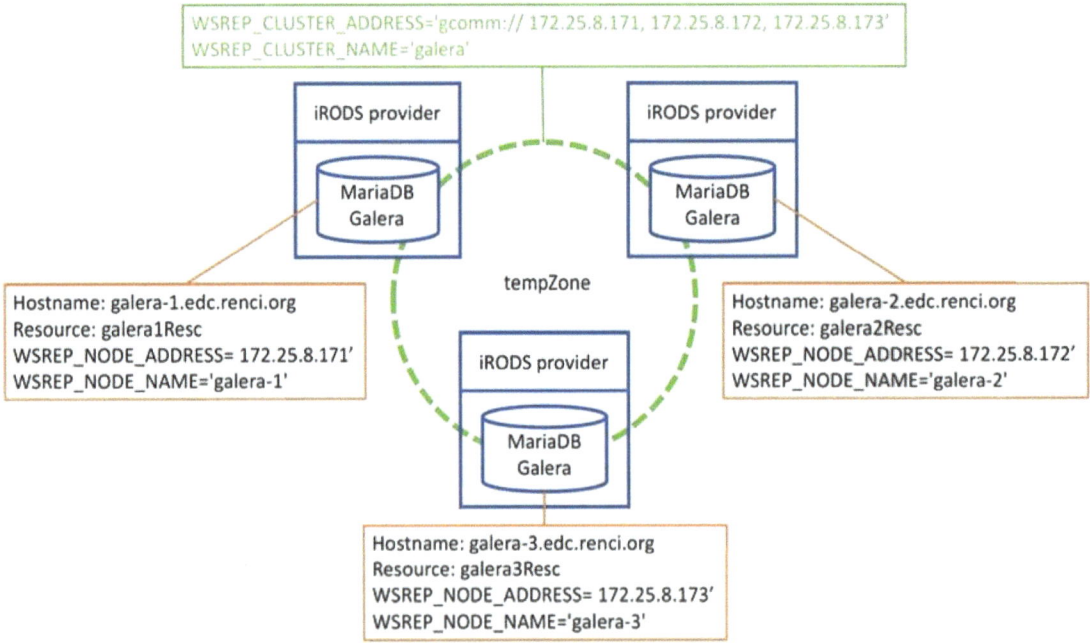

Figure 1. Baseline Docker deployment of three iRODS Catalog Providers via MariaDB Galera Cluster.

The iRODS provider node[6][7] is using MariaDB Galera cluster in Docker[8] and is based on a centos:7 image[9].

Design

The proof of concept has been designed to run in Docker and does not prohibit storing the files for iRODS and MariaDB to the host. Generally an iRODS system deployment would use service level accounts named **irods** and **mysql** to retain/own the iRODS service files and the MariaDB files respectively. The Docker image used herein has defaulted these system users to be:

```
irods: UID=996, GID=996
mysql: UID=997, GID=997
```

If the user chooses to deploy the **irods-provider-galera** image using local volume mounts, then these **UID** and **GID** values would be found on the local system for the **user:group** in charge of the shared volumes. The **UID** and **GID** for both the **irods** and **mysql** user can be set at runtime to be any valid combination within the host system the container is being run from.

Setup

A set of directories in `/var` are set aside to mount to the **irods-provider-galera** container to hold the iRODS and MariaDB files. The following `setup-galera` script outlines the creation and initial permissions for these directories:

```bash
#!/usr/bin/env bash
sudo rm -rf /var/galera
sudo mkdir -p /var/galera/init
sudo mkdir -p /var/galera/vault
sudo mkdir -p /var/galera/var_mysql
sudo mkdir -p /var/galera/var_irods
sudo mkdir -p /var/galera/etc_irods
sudo chown -R galera:galera /var/galera
exit 0
```

The user named **galera** will be in charge of running **irods-provider-galera**, and has the following attributes:

```
$ id galera
uid=2112(galera) gid=2112(galera) groups=2112(galera),992(docker)
```

When instantiating the iRODS container we set the environment variables of the **irods** and **mysql** users to use UIDs and GIDs that are locally available:

```
-e UID_MYSQL=2000 \ # UID that is unassigned on the localhost
-e GID_MYSQL=2112 \ # GID assigned to user galera on the localhost
-e UID_IRODS=2112 \ # UID assigned to user galera on the localhost
-e GID_IRODS=2112 \ # GID assigned to user galera on the localhost
```

View of `/var/galera` before **irods-provider-galera** is run:

```
$ ls -alh /var/galera/
total 4.0K
drwxr-xr-x   7 galera galera   77 Jun 11 11:34 .
drwxr-xr-x. 21 root   root   4.0K Jun 11 11:34 ..
drwxr-xr-x   2 galera galera    6 Jun 11 11:34 etc_irods
drwxr-xr-x   2 galera galera    6 Jun 11 11:34 init
drwxr-xr-x   2 galera galera    6 Jun 11 11:34 var_irods
drwxr-xr-x   2 galera galera    6 Jun 11 11:34 var_mysql
drwxr-xr-x   2 galera galera    6 Jun 11 11:34 vault
```

The next three sections will step through testing the clustered iCAT, first with Docker deploying all three nodes on a single machine, then Docker deploying an iRODS catalog provider on three VMs, and finally on three VMs with an additional latency introduced to simulate longer distances.

LAN - LOCAL MACHINE

The first test runs on a single VM and was created to demonstrate the basic principles of using the MariaDB Galera cluster as the iRODS catalog for multiple provider nodes. The test script does the following:

- Creates a local Docker network named `galeranet` so that known IP addresses can be assigned to each node.
- Stands up the initial bootstrap node using mostly defaults as set by the Docker image.
- Stands up two additional nodes in series that join the cluster named `galera` as they discover others on the local `galeranet` network.

As each node completes its stand up routine, it reports the number of nodes participating as the `wsrep_cluster_size`, lists the databases and the grants for user `'irods'@'localhost'`, and finally prints out all tables within the iCAT database.

Since this initial test was performed on a single VM using a Docker network, it was not subjected to any external testing and only subject to simple iCommands to validate synchronization between nodes and partitioning between named node resource definitions.

The expected output was observed and showed that the three nodes were up and connected within the Docker network:

```
1  $ docker ps
2  CONTAINER ID   IMAGE                                  STATUS          NAMES
3  b51fcb1501ec   mjstealey/irods-provider-galera:4.2.1  Up 44 minutes   irods-galera-node-3
4  654b094fc554   mjstealey/irods-provider-galera:4.2.1  Up 45 minutes   irods-galera-node-2
5  cc51c3413306   mjstealey/irods-provider-galera:4.2.1  Up 45 minutes   irods-galera-node-1
```

LAN - VIRTUAL MACHINES

The second test runs on three VMs with Docker running a clustered iRODS catalog provider on each and uses the built-in iRODS test suite located at `/var/lib/irods/scripts/run_tests.py`. The limitation of the test suite is that it was not necessarily designed to run against a clustered database, so the default notion of `demoResc` is problematic when the goal is to test across a distributed iCAT cluster. Because of this limitation, we issued `python run_tests.py --run_python_suite` on each VM/node within the cluster, one at a time.

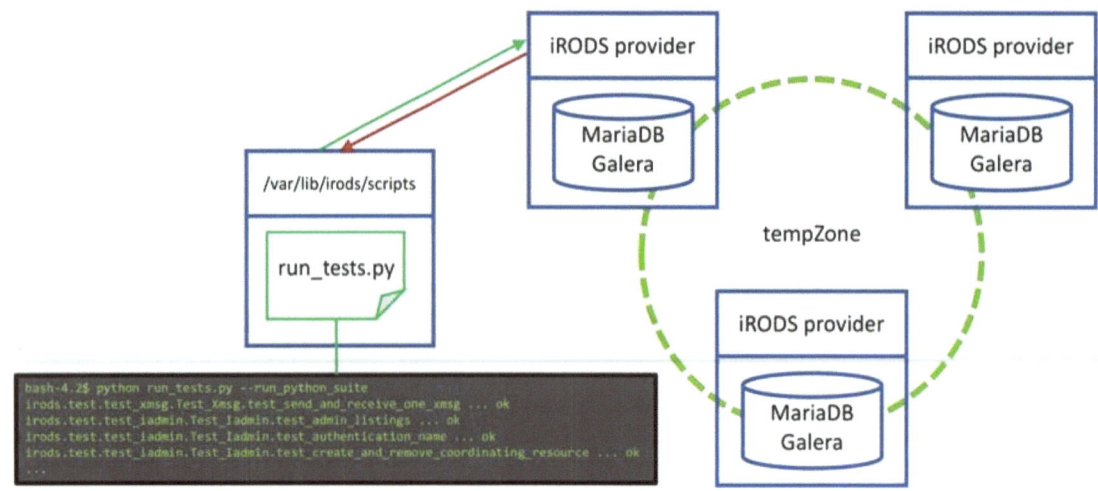

Figure 2. Simple test run on one of the three nodes in a clustered catalog provider.

While the test suite runs on one node, the other nodes were manually monitored via various iCommands and SQL queries validating that the test files and corresponding database entries were being created as expected and visible from the other nodes.

The `run_tests.py` script is the default script used for iRODS testing and is part of a normal iRODS installation. As an initial check, it was chosen to run with the `--run_python_suite` option which takes roughly four hours to complete all its nearly 1500 tests.

Since iRODS is being run in Docker, it is necessary to invoke the test suite from within the Docker container. This can be accomplished by connecting to the container as the **irods** user, changing to the `scripts` directory, and issuing the test run call.

```
1  [galera@galera-1 ~]$ docker exec -ti -u irods irods-galera-1 /bin/bash
2  bash-4.2$ cd ~
3  bash-4.2$ pwd
4  /var/lib/irods
5  bash-4.2$ cd scripts/
6  bash-4.2$ python run_tests.py --run_python_suite
7  irods.test.test_xmsg.Test_Xmsg.test_send_and_receive_one_xmsg ... ok
8  irods.test.test_iadmin.Test_Iadmin.test_addchildtoresc_forbidden_characters_3449 ... ok
9  irods.test.test_iadmin.Test_Iadmin.test_admin_listings ... ok
10 irods.test.test_iadmin.Test_Iadmin.test_authentication_name ... ok
11 irods.test.test_iadmin.Test_Iadmin.test_create_and_remove_coordinating_resource ... ok
12 ...
```

The running tests can be observed from the other nodes via iCommands or mysql queries (in this case, `ils ../` from `irods-galera-2`:

```
1  [galera@galera-2 ~]$ docker exec -ti -u irods irods-galera-2 ils ../
2  /tempZone/home:
3    C- /tempZone/home/alice
4    C- /tempZone/home/bobby
5    C- /tempZone/home/issue_3104_user
6    C- /tempZone/home/otherrods
7    C- /tempZone/home/public
8    C- /tempZone/home/rods
```

Over nearly twelve hours, the three test runs completed, in turn, from each node in the cluster and the outputs were consistent with running the test suite in a single server configuration.

```
1  $ python run_tests.py --run_python_suite
2  irods.test.test_xmsg.Test_Xmsg.test_send_and_receive_one_xmsg ... ok
3  irods.test.test_iadmin.Test_Iadmin.test_addchildtoresc_forbidden_characters_3449 ... ok
4  irods.test.test_iadmin.Test_Iadmin.test_admin_listings ... ok
5  irods.test.test_iadmin.Test_Iadmin.test_authentication_name ... ok
6  irods.test.test_iadmin.Test_Iadmin.test_create_and_remove_coordinating_resource ... ok
7  ...
8  irods.test.test_irmdir.Test_Irmdir.test_irmdir_of_collection_containing_dataobj ... ok
9  irods.test.test_irmdir.Test_Irmdir.test_irmdir_of_dataobj ... ok
10 irods.test.test_irmdir.Test_Irmdir.test_irmdir_of_empty_collection ... ok
11 irods.test.test_irmdir.Test_Irmdir.test_irmdir_of_nonexistent_collection ... ok
12 irods.test.test_iquest.Test_Iquest.test_iquest_MAX_SQL_ROWS_results__3262 ... ok
13
14 ----------------------------------------------------------------------
15 Ran 1468 tests in 13106.840s
16
17 OK (skipped=89)
18 <__main__.RegisteredTestResult run=1468 errors=0 failures=0>
```

WAN - VIRTUAL MACHINES

The third test was designed to run a parallel test script that puts and gets large files into iRODS in a continuous loop until broken by the user. This script uses GNU parallel[10].

The parallel put/get test was first run on the testbed without any additional latency introduced to the nodes and allowed to run for six hours without error.

Then, varying amounts of latency were added to each node using NetEm to simulate the distances involved with real world WAN deployments of iRODS. The parallel test script was again allowed to run against the nodes in this configuration for multiple hours.

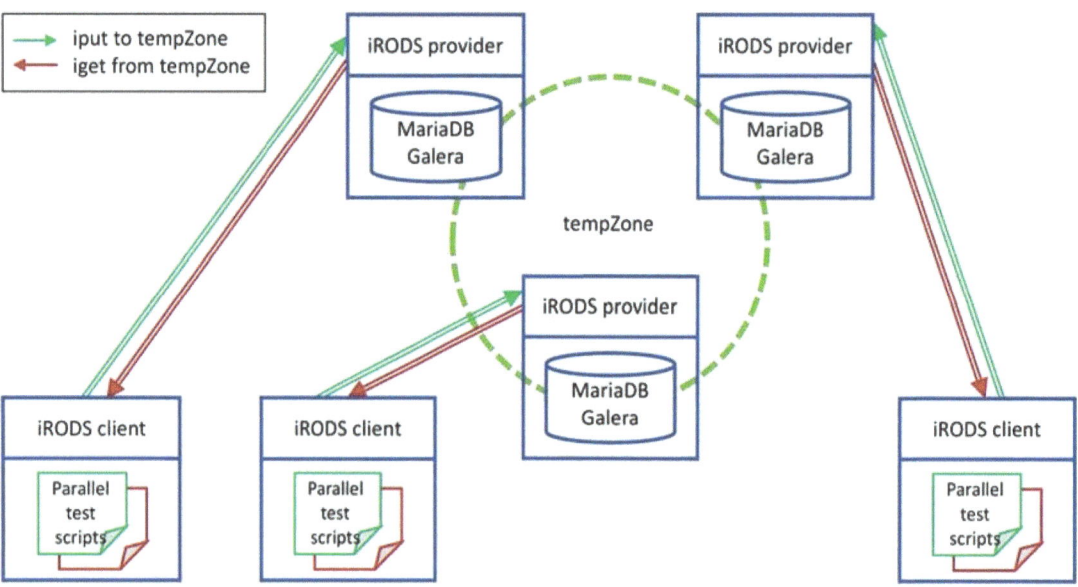

Figure 3. Concurrent parallel test scripts running on a three-node clustered catalog provider.

The following real-world latency approximations were used to add delay to each node, respectively:

Low National (Chicago/RENCI):	20ms
Medium National (Coast to Coast - SanFrancisco/RENCI):	60ms
International (RENCI/Netherlands):	117ms

The approximations were implemented via NetEm as follows:

```
1  galera-1: sudo tc qdisc add dev eth0 root netem delay 20.0ms
2  galera-2: sudo tc qdisc add dev eth0 root netem delay 60.0ms
3  galera-3: sudo tc qdisc add dev eth0 root netem delay 120.0ms
```

Since NetEm affects both incoming and outgoing traffic, the effect was cumulative across nodes which can be observed in the RTT between nodes.

	galera-1	galera-2	galera-3
galera-1	n/a	80.5ms	140.5ms
galera-2	80.5ms	n/a	180.4ms
galera-3	140.5ms	180.5ms	n/a

Table 1. Round Trip Time (RTT) within the three-node testbed

After the initial parallel test on a single node, the `parallel_get_put.sh` script was launched simultaneously on all three nodes. The scripts first generate 256 40 MB files and then start sending them via `iput` in 30 parallel threads to the local iRODS server instance onto its local storage resource. Once all 256 files have been transferred, they are retrieved using `iget` in 30 parallel threads. This process is embedded in a loop and continues until an error is encountered or is manually stopped.

The three nodes ran without error for over two hours until manually interrupted and forcibly quit. During this period each node completed a number of put/get loops based on the distance from its peers:

galera-1	48 loops
galera-2	12 loops
galera-3	3 loops

There was a single multi-master deadlock error during this testing where the database complained of a timeout during the concurrent writes. This was due to the relatively large number of operations that occur within the database transaction during a single iRODS connection. We have since heard anecdotal evidence of this same case from others; concurrent load tripping a cluster timeout when the latency is high between cluster nodes. The clustered nodes have a default timeout setting that interprets a delay as their peer not responding. We expect that adjusting the default timeout can reduce the rate of incidence.

In addition to adjusting the timeout, there are two main ways that iRODS can improve to avoid deadlock errors of this type. The first would be to add a simple retry upon network timeout. Such a retry would push back the threshold where the concurrent writes trigger a timeout, but ultimately, would be a temporary fix and would not solve the problem for a growing system. A more robust path would be to reduce the number of operations that occur within the database transactions themselves. If the database is doing less work within each transaction, it is less likely that the system will hit a timeout and interpret it as a deadlock.

FUTURE WORK

This exercise has demonstrated that a distributed cluster for the iCAT is feasible, but that there is much more work to do. The most straightforward efforts will be to spend more time testing on real networks over real distances, rather than simulated. A second focus should be to test other database technologies besides MariaDB's Galera. We are interested in working with CockroachDB as the next viable candidate. The third type of future work should be testing targeted edge cases that will push on the overlap and potential friction between the database itself and how iRODS aims to present a consistent surface to its clients. And lastly, we need to investigate the upgrade or migration path for how an existing iRODS Zone moves from a singular iCAT to a clustered iCAT.

CONCLUSION

iRODS works when using a clustered SQL database and can satisfactorily address the use case defined by global organizations that want to leverage a unified namespace. Like with all new use cases, some defaults settings should be adjusted to accommodate a different set of assumptions. The iRODS architecture proved flexible enough to easily adapt to this new use case of a high latency, WAN, multi-master deployment, however we need more testing and real world use cases to drive the next set of improvements.

ACKNOWLEDGEMENTS

The authors would like to thank the ongoing support of the Renaissance Computing Institute (RENCI) at UNC-Chapel Hill for the networking and virtual machine infrastructure to conduct these tests. In addition, the expertise and use case support from the SciDAS[11] project and BioTeam[12] proved invaluable.

REFERENCES

[1] iRODS website. https://irods.org
[2] iRODS. https://github.com/irods/irods
[3] iRODS Primer 2: Integrated Rule-Oriented Data System (2017). Hao Xu, Terrell Russell, Jason Coposky, Arcot Rajasekar, Reagan Moore, Antoine de Torcy, Michael Wan, Wayne Schroeder, and Sheau-Yen Chen. Synthesis Lectures on Information Concepts, Retrieval, and Services, March 2017, Vol. 9(3):1-131. https://doi.org/10.2200/S00760ED1V01Y201702ICR057
[4] MariaDB website. https://mariadb.org/
[5] Stephen Hemminger (2005). NetEm - Emulating Real Networks in the Lab. . Linux Conf, Canberra, Australia. April 2005. https://linux.org.au/conf/2005/Papers/Stephen%20Hemminger/index.html
[6] Michael Stealey (2017). iRODS Provider, Galera. https://github.com/mjstealey/irods-provider-galera
[7] Michael Stealey (2017). Documentation for iRODS Provider, Galera. https://mjstealey.github.io/irods-provider-galera/
[8] Michael Stealey (2017). MariaDB Galera Docker Container. https://github.com/mjstealey/mariadb-galera
[9] Docker Hub - CentOS. https://hub.docker.com/_/centos/
[10] O. Tange (2011). GNU Parallel - The Command-Line Power Tool. login: The USENIX Magazine, February 2011:42-47.
[11] Frank Feltus, Claris Castillo, Ray Idaszak, Melissa Smith, Stephen Ficklin (2017). National Cyberinfrastructure for Scientific Data Analysis at Scale (SciDAS). NSF Award 1659300. https://www.nsf.gov/awardsearch/showAward?AWD_ID=1659300
[12] BioTeam website. https://bioteam.net/

FAIR Sequencing Data Repository based on iRODS

Felipe O. Gutierrez, Paul De Geest, Aldo Jongejan, Sjoerd Repping, J.T. van den Berg, Antoine H.C. van Kampen, Sílvia D. Olabarriaga
Academic Medical Center of the University of Amsterdam - Amsterdam, NL
{f.oliveiragutierrez, p.f.degeest, a.jongejan, s.repping, j.t.vandenberg, a.h.vankampen
s.d.olabarriaga}@amc.uva.nl

Diogo F.C. Patrão
A. C. Camargo Cancer Center - São Paulo, Brazil
djogopatrao@gmail.com

ABSTRACT

Research data management (RDM) and the FAIR principles (Findable, Accessible, Interoperable, Reusable) are widely promoted as basis for a shared research data infrastructure. Nevertheless, researchers involved in next generation sequencing (NGS) still lack adequate RDM solutions. The NGS metadata is generally not stored together with the raw NGS data, but kept by individual researchers in separate files. This situation complicates RDM practice. Moreover, the (meta)data does often not meet the FAIR principles [6]. Consequently, a central FAIR-compliant repository is highly desirable to support NGS related research. We have selected iRODS (Rule-Oriented Data management systems) [3] as a basis for implementing a sequencing data repository because it allows storing both data and metadata together. iRODS serves as scalable middleware to access different storage facilities in a centralized and virtualized way, and supports different types of clients. This repository will be part of an ecosystem of RDM solutions that cover complementary phases of the research data life cycle in our organization (Academic Medical Center of the University of Amsterdam). We selected Virtuoso [5] to enrich the metadata from iRODS to enable the management of a triplestore for linked data. The metadata in the *iCat* (iRODS' metadata catalogue) and the ontology in Virtuoso are kept synchronized by enforcement of strict data manipulation policies. We have implemented a prototype to preserve raw sequencing data for one research group. Three iRODS client interfaces are used for different purposes: *Davrods* [4] for data and metadata ingestion, data retrieval; *Metalnx-web* [7] for administration, data curation, and repository browsing; and *iCommands* [2] for all tasks by advanced users. Different user profiles are defined (principal investigator, data curator, repository administrator), with different access rights. New data is ingested by copying raw sequence files and the corresponding metadata file (a *sample sheet*) to the *landing* collection on iRODS. An iRODS rule is triggered by the sample sheet file, which extracts the metadata and registers it to the iCAT as *AVU* (Attribute, Value and Unit). Ontology files are registered into Virtuoso. The sequence files are copied to the *persistent* collection and are made uniquely identifiable based on metadata. All the steps are recorded into a report file that enables monitoring and tracking of progress and faults. Here we describe the design and implementation of the prototype, and discuss the first assessment results. Initial results indicate that the proposed solution is acceptable and fits the researchers workflow well.

Keywords

Next Generation Sequencing, Research Data Management, iRODS, FAIR principles, genomics data, ontology.

INTRODUCTION

As part of modern biomedical research many types of (large) data sets are produced. OMICS experiments - and in particular next generation sequencing (NGS) - are no exception. Biomedical research involving NGS data generally comprises collaborations between clinical departments, laboratories, and data analysis groups that have different cultures and procedures for working with data. With the growth of data, it has become challenging to keep track of data, processes and outcomes of research over long periods of time and across the collaborating units. Adequate

management of research data has become paramount to guaranteeing efficiency and integrity of research, as well as to enable future data reuse and exploitation. Important initiatives concerning OMICS data management are supported by, for example, the BBMRI and ELIXIR programs which are active across Europe. Various of these initiatives promote and highlight the FAIR principles for research data [6]. According to these principles, research data should be Findable, Accessible, Interoperable and Reusable, which are properties that demand great care in terms of collection, annotation and archival. Nevertheless, few solutions exist for NGS data management that accomodate these principles. In practice, often there is no central repository for sequencing data, and each individual researcher is responsible for the storage and backup of their data and metadata. Moreover, NGS related metadata is generally not generated automatically, requiring manual annotation by the researchers. And finally, metadata is typically not stored together with the NGS data.

In our prototype we aim to develop a repository for raw NGS research data for our organization. The repository should adhere to the FAIR principles by defining standardized, minimum, interoperable metadata which needs to be preserved together with the data. iRODS [3] is used as a basis for this solution, in combination with a variety of clients to fulfill the diverse needs of the different user roles. This paper briefly introduces characteristics of NGS data before describing the designed solution. Preliminary results of a pilot case study are presented and discussed.

NGS RESEARCH DATA REPOSITORY

NGS data is generated by sequencing the genetic material within biological samples i.e. measuring the genetic material and thereby defining the order of its nucleotides. An NGS experiment generally comprises the following steps:

- prepared biological samples are sent to a sequencing facility;
- the sequencing facility performs one or more sequencing runs using a sequencing instrument (sequencer);
- the output of these runs are images that are converted to raw sequencing data (this data comprises several million reads). The NGS data is stored in a standard file format, e.g.: **fastq** (FASTQ format is a text-based format for storing both a biological sequence (usually nucleotide sequence) and its corresponding quality scores.).
- information (metadata) about the run(s), sample(s) and output file(s) are described in the so called **sample sheet**, which can take different formats and contain different contents across sequencing platforms and services.

NGS data of such form is being generated in our organization by multiple research groups that, at the moment, need to take care of their own data storage and management necessities. Metadata related to these data are being maintained locally by the respective researchers. Data and metadata are spread in various systems, using non-standardized protocols, making data retrieval, traceability and reuse extremely difficult. Lastly, due to being an academic hospital the (meta)data can be highly sensitive and therefore can not leave the site, excluding any external repositories from being used. In our project we aim to design a repository that can be used by various research groups in our organization which would also be a viable option for similar organizations. In order to follow the FAIR principles we aim to harmonize the format and content of the sample sheets, to represent the metadata in an ontology, and to store the metadata and data together. Note that the metadata associated with a raw NGS dataset is potentially extremely rich, describing the whole process from the biological sample acquisition through the wet lab sample preparation and finally sequencing. Collecting such detailed metadata is not a trivial task for researchers. Different research groups have their own priorities for NGS related metadata, therefore, in this project, we allow each research group to define a subset of the recommended metadata fields, in addition to the minimum of required metadata for an NGS experiment. This flexibility is intended to make the system more user friendly and increase the willingness of the researcher to provide this metadata, which comes at the expense of standardization.

REPOSITORY ARCHITECTURE

The main components of the proposed system are presented in Figure 1: iRODS as the main data and metadata storage server, Virtuoso as the ontology server, the different access points and connection protocols. Openlink

Virtuoso [5] is a hybrid database engine, allowing not only the management of relational databases but also of a *triplestore* for linked data. It provides support for all major ontology formats and allows data-access through the Virtuoso conductor web-interface.

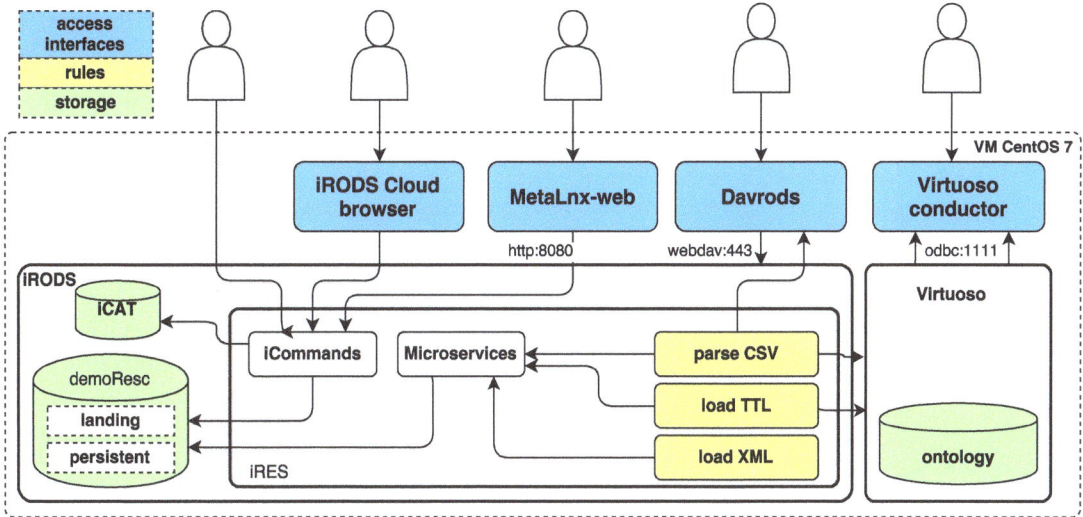

Figure 1. Main components of our FAIR NGS Data repository: iRODS for files and metadata, Virtuoso for RDF store, four types of user interfaces (blue), and rules (yellow). Data and metadata storage resources are depicted in green. A data repository is a collection with two areas called landing (for data ingestion) and persistent (for retrieval). Rules are triggered by new data on the landing area, and use microservices to create and load metadata and ontology.

Within iRODS there are currently two storage resources (the iCat and the data storage space) which are placed in the same storage pool. The data storage space is further divided into repositories that correspond to collections that are potentially owned by different research groups. Each repository has two directories, each with different access types or permissions (*landing* and *persistent*). The landing directory is temporarily used for new data and metadata ingestion and the persistent directory holds data and metadata permanently. Rules in iRODS are used for extracting metadata from files written into the landing directory, loading them into the iCAT and Virtuoso [5], and moving them to the persistent directory.

The users can directly access iRODS in three ways, depending on their needs and knowledge: iCommands [2], Davrods [4] and MetaLnx-web [7]. Davrods [4] is an Apache WebDAV interface that provide access to iRODS. It is a bridge between the WebDAV protocol and the iRODS API, implemented as an Apache HTTPD module. It leverages the Apache server implementation of the WebDAV protocol, mod_dav, for compliance with the WebDAV Class 2 standard. Through Davrods the user can ingest and retrieve data from iRODS, mounting it as a file system in his/her workstation. Therefore, the user cannot work with metadata through this client interface. Metalnx-web [7] is a web application designed to work alongside iRODS. It provides a graphical UI that can help simplify most administration, collection management, and metadata management tasks removing the need to memorize the long list of iCommands. This is the most rich interface to access iRODS, it includes functionalities to execute all iCommands using a graphical interface and also retrieve data and metadata.

FUNCTIONAL WORKFLOW

Figure 2 presents the functional workflow. It describes a generic approach for ingesting data into iRODS (through Davrods), for generating metadata into iRODS and ontology into Virtuoso (through iRODS rules), and for retrieving data and metadata.

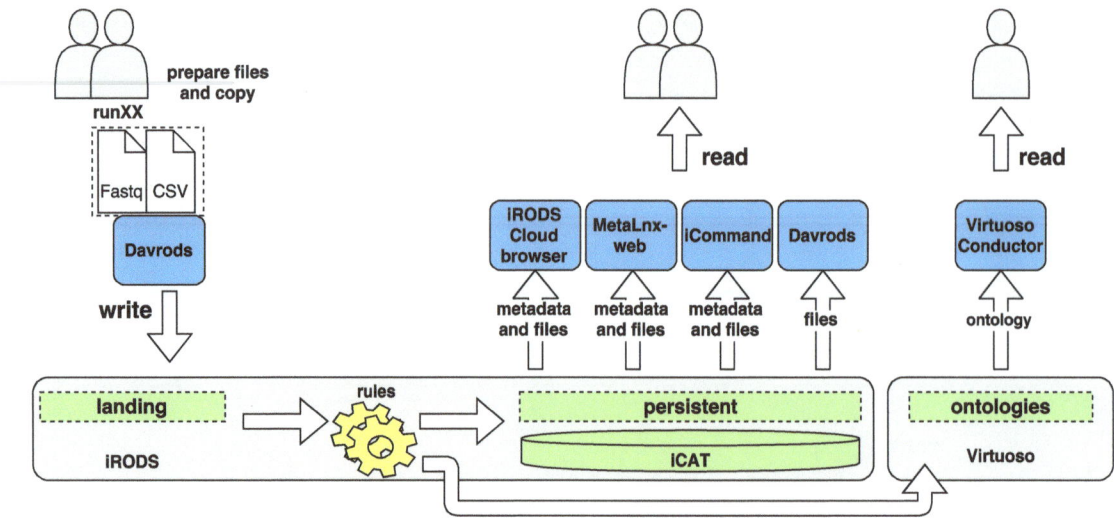

Figure 2. Functional workflow to ingest and retrieve (meta)data from iRODS. Downwards arrows indicate data/metadata ingestion flow, upwards arrows indicate retrieval flows and the horizontal arrows indicate automated processing.

Before ingestion to the repository, the sequencing files and the metadata need to be collected into a single directory that has the name of the sequence run that will be ingested into iRODS. This directory is represented by the name **runXX** in Figure 2, and it contains all fastq files and one sample sheet (CSV file). Once the user is logged in through a mounted Davrods client, they can copy the complete directory to the landing directory. The rules are triggered when the sample sheet, a CSV file, is copied through Davrods into a landing directory of an iRODS collection and the sample sheet is correctly interpreted. Each rule has its specific set of functions and they use several microservices from iRODS. At the end of the rule execution, all the results are moved to the persistent directory, which has read-only access and the metadata are loaded into iRODS and Virtuoso, allowing the users to query and find their data in an easy way, improving Findability (see FAIR principles).

The rules implement several steps and each step is monitored along the workflow and possible failures are reported into a *report file*. This report is stored together with the data, and it can be consulted by the user to check if the data and metadata were loaded correctly into the repository. Each line of the report file matches to one step in the workflow, indicating also the date, time, the path to the file that was being handled and the resulting status. If the rule executed correctly, the final result is *OK*; otherwise, an error message explains the specific step that went wrong. This report file is encoded in *html*, for a user friendly view of all steps and messages.

USER POLICIES AND ROLES

Different research groups need to have separated repositories for their NGS research data. We create iRODS collections, users and groups to organize them inside the storage server. One of the policies of our iRODS instance is for creating metadata in iCat and Virtuoso automatically when the researchers ingest data (see figure 1) on their own landing directories. additionally, we can define different rules for each research group. Furthermore, the two different directories at the storage space have different permission policies. On the landing directory the user has permissions to write files, and on the persistent directory the user has read-only permissions. We have enabled the SSL connection on iRODS server, so all clients that connect to it are using a secure protocol. The iRODS users have the same access rights through Metalnx-web as on iCommands. The Davrods access works with the group access, so all users inside a specific group can have access to specific *landing* and *persistent* directories.

We have defined four principal user roles for our system. The most basic user role is the *Researcher* that only has access to the landing and persistent storage places for the groups they belong to. The files ingested into the landing directory are private to this user. The files and metadata in the persistent directory are readable by all users in the group. The *Data Steward* has full access to the user group directory they belong to, including rights to change the metadata. The *Principal Investigator* has full control of the repository, being able to include users into groups and assign roles. Finally, the *Repository Administrator* has the ability to create rules and manage all services in the system. The three first users belong mainly to iRODS, whereas the last user also has control of the server.

USE CASE EVALUATION

The use case of a single research group is currently used for evaluating the prototype environment. A repository was created for this group with corresponding collection, group and user accounts. A user was assigned by the research group and trained to use the system. The goal was to store some of the existing raw NSG data into the repository with minimum metadata. The datasets used in this evaluation covered varied conditions regarding the number and size of files. The beta user connected to the repository from a Linux desktop using a mounted Davrods drive for data ingestion and retrieval, and iCommands for metadata search. The metadata and data preparation was supported by the *Data Steward* from our team for the first datasets. The user, *Data Steward* and developer had close communication during the whole process to discuss difficulties and design improvements which were then implemented using the agile development methodology [1]. We monitored and tested the user experience along the phases, detailed in the workflow section, and summarized here:

- **Data and metadata preparation:** the user prepares a directory with NGS data files and a sample sheet (CSV file) with metadata that follows a specific format.
- **Data and metadata ingestion:** the user accesses the Davrods mount point drive to copy data and metadata to the landing directory.
- **Rule processing:** the sample sheet triggers an iRODS rule, generating a XML file with AVU metadata (Attribute, Value, Unit) to be ingested into iRODS and a RDF file to be ingested into Virtuoso.
- **Data and metadata retrieval:** Metalnx and iCommands are used to search files through metadata. Davrods can be used to retrieve data only.

We do this monitoring by asking the user directed questions regarding the following topics:

- **clarity for the user:** does the user know what to do in each case? For example, we wish to understand whether the sample sheet requirements were clear enough to enable successful data ingestion for varied datasets.
- **feasibility for the user:** is it feasible for the user to accomplish the requirements in a practical and workable manner?
- **ease of use:** is it easy for the user to use the system? ie.: mounting the landing directory, copying files, etc.
- **feedback for the user:** is the feedback coming from the system sufficient, clear, easy to find and timely?
- **robustness:** to determine if the system performs well, not only under ordinary conditions, but also under unusual conditions that stress its designers assumptions.
- **performance with different workloads:** to determine if the system performs well with more than one user accessing it through different computers, as well with different sets of files.

Below we briefly describe our preliminary results for this usability test in a qualitative manner. We found that during preparation of the data and metadata the user had only minor problems with user input in the sample sheet and file names, e.g. the user left certain mandatory fields open. The user was mostly satisfied with the process of ingesting the data and metadata after mounting the Davrods interface on their personal computer, although it was noted that due to the amount of data files being ingested the system can become slower. Additionally, the user had some minor comments regarding the level of detail of the feedback coming from the system. In terms of rule processing the user experienced some more serious problems, e.g.: mac OS systems generate hidden files that caused errors on the system side. As for retrieving data and metadata the user experienced minor problems learning iCommands but were mostly satisfied using Metalnx-web, which implements all iCommands available.

DISCUSSION

Metadata is recognized as useful on all types of RDM system, however annotating detailed metadata takes much effort. Therefor, we opted for a simple approach that is feasible for the users. The simplicity of the current sample sheet, which is mostly generated already by the sequencing facilities, has been considered an advantage during this first evaluation. The choice for iRODS as a basis for our NGS repository has also been considered positive in this first evaluation. It proved to be flexible to accommodate various changes, in particular regarding metadata ingestion, which is expected to be a dynamic component of any RDM system that evolves along time. We expect that this technology will become better supported in our organization, leading to long term sustainability for the solution proposed here.

Last, and most important, is the usability of the system. We have set up the prototype in a way that the user doesn't need to have any iRODS knowledge by: choosing to use Davrods allowing users to drag and drop files to the system; processing the metadata automatically using iRules; using Metalnx-web as an interface for retrieving data and metadata; having a *data steward* responsible for helping successful ingestion of the data and metadata; and having a *administrator* responsible for implementing new policies through iRules. We believe that these choices make our prototype user friendly.

Based on the preliminary results of our usability tests we found that the prototype was generally a feasible solution for the management of NGS data and metadata and we could reach the goals of our prototype even though it is very simple for this first version. Additionally, we found that performance improvements of the *landing* directory could be made. Finally, since we chose to reuse existing generic clients to access the repository, we observed that they provide only limited contextual feedback. Therefore more user-friendly tools that are better customized to this application area might be necessary to address the needs of other user profiles in the future.

ACKNOWLEDGMENTS

This project is financially supported by the AMC Innovation Fund. The authors would like to thank Rudy Scholte, Lieuwe Kool, Joyce Nijkamp, Barbera van Schaik and Niek de Vries for their insightful contributions that helped shape up our RDM solution.

REFERENCES

[1] Agile methodology. `http://agilemethodology.org/`. Accessed: 2017-05-01.

[2] irods icommands. `https://docs.irods.org/4.1.10/`. Accessed: 2017-05-01.

[3] A. Rajasekar, R. Moore, C.-y. Hou, C. A. Lee, R. Marciano, A. de Torcy, M. Wan, W. Schroeder, S.-Y. Chen, L. Gilbert, et al. irods primer: integrated rule-oriented data system. *Synthesis Lectures on Information Concepts, Retrieval, and Services*, 2(1):1–143, 2010.

[4] T. Smeele and C. Smeele. Davrods - an Apache WebDAV interface to iRODS. *iRODS User Group Meeting 2016 Proceedings*, 1:41–47, Dec. 2016.

[5] Openlink Virtuoso software. `https://virtuoso.openlinksw.com/`. Accessed: 2017-05-01.

[6] M. D. Wilkinson et al. The FAIR guiding principles for scientific data management and stewardship. *Scientific data*, 3, 2016.

[7] S. Worth. An administrative and metadata ui for irods. `https://github.com/Metalnx/metalnx-web`. Accessed: 2017-05-01.

Swedish National Storage Infrastructure for Academic Research with iRODS

Ilari Korhonen
KTH Royal Institute of Technology
SE-100 44 Stockholm, Sweden
ilarik@kth.se

Dejan Vitlacil
KTH Royal Institute of Technology
SE-100 44 Stockholm, Sweden
vitlacil@kth.se

Janos Nagy
Linköping University
SE-581 83 Linköping, Sweden
fconagy@nsc.liu.se

Krishnaveni Chitrapu
Linköping University
SE-581 83 Linköping, Sweden
krishnaveni@nsc.liu.se

Ilker Manap
KTH Royal Institute of Technology
SE-100 44 Stockholm, Sweden
manap@kth.se

ABSTRACT

The Swedish National Infrastructure for Computing (SNIC) has decided to invest in a large scale distributed iRODS-based storage infrastructure for complementing its national storage offering for academic research. Currently the SNIC Swestore national storage service is relying on dCache as its storage solution while it has grown into a petascale operation. However, many users and research groups have expressed interest in different access methods and functionalities (such as the use of multiple different user authentication methods and metadata management) than what can be easily accommodated with dCache. This prompted the investigation of extending the national storage service with the use of different storage technologies. A project was commenced by SNIC for an iRODS-based distributed scalable storage service complementing Swestore. The SNIC iRODS project has now been concluded and the resulting system is being installed into a production environment and integrations are being set in place with other SNIC services. Our accomplishments including but not limited to: a model for the deployment of a geo-replicated iRODS iCAT over two administrative domains with a DNS-based failover mechanism; a model for the deployment of existing and future distributed storage resources within SNIC with iRODS; a novel iRODS interface for tape resources written against the IBM Spectrum Protect (TSM) API; improvement for iRODS logging capabilities with a syslog forwarder; contributions for iRODS user authentication via an alternative PAM authenticator; automated provisioning of iRODS grids and associated services with Ansible, optionally provisioning clusters of VM's with Vagrant for testing; integration of the SNIC iRODS storage service with the SNIC User and Project Repository (SUPR) for provisioning of iRODS users and groups for approved proposals. This solution enables the easy integration of local HPC storage solutions as well as EUDAT, which delivers data to the HPC, HTC and cloud services in Europe.

Keywords

Research data, metadata management, infrastructure.

INTRODUCTION

Swedish National Infrastructure for Computing (SNIC) is a national initiative in Sweden responsible for the financing of most of the High Performance Computing (HPC) and related data storage activities in Sweden. SNIC is being funded by Vetenskapsrådet (Science Council), which itself is a governmental agency in Sweden, tasked to guarantee a high level of research in all fields of science.

iRODS UGM 2017 June 13-15, 2017, Utrecht, Netherlands
[Authors retain copyright.]

SNIC distributes funding between the major Swedish HPC centers, of which currently there are six: PDC in KTH Royal Institute of Technology (Stockholm), NSC in Linköping University (Linköping), UPPMAX in Uppsala University (Uppsala), LUNARC in Lund University (Lund), C3SE in Chalmers University (Gothenburg) and HPC2N in Umeå University (Umeå).

SWEDISH NATIONAL STORAGE INFRASTRUCTURE

SNIC coordinates and provides high end computing and storage capacity for Swedish academic research and education. For this purpose, SNIC provides a set of resources to meet the needs of researchers from all scientific disciplines and from all Swedish universities, university colleges and research institutes. Swestore is a National Research Data Storage Infrastructure operated by SNIC.

The resources provided by Swestore are made available through open procedures such that the best Swedish research is supported and new research is facilitated. Prioritization and allocation of resources must be done in a clear and transparent manner, based on scientific quality, scientific need and technical feasibility of using the requested resources efficiently.

The purpose of Swestore allocations, granted by Swedish National Allocations Committee (SNAC), is to provide large scale data storage for 'live' or 'working' research data, also known as active research data.

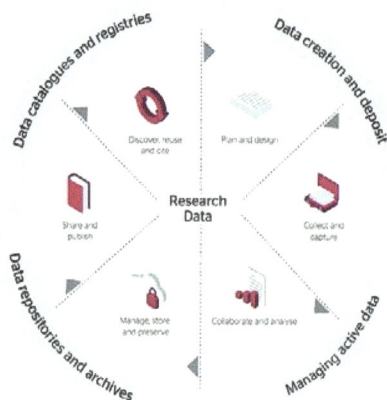

Figure 1. Research Data Lifecycle (©Jisc and Bonner McHardy [CC BY-NC-ND])

Individuals and groups, eligible to apply, can apply for allocations on the available SNIC resources. The entire process is managed by SUPR (SNIC User and Project Repository), which is the self-service portal for users at all SNIC centres. It is the SNIC database used to keep track of persons, projects, project proposals and more.

Once an application has been approved, a project will be established on the resources on which allocations are granted and users are informed. Initial authentication and authorisations are set accordingly. The applicant will be responsible (PI) for the project and must keep an overview of the usage of the allocation(s).

When the research project is over the SNIC users are required to remove their research data to more suitable data services and resources (e.g long-term preservation and archiving, data repositories, data catalogues and registries).

For the past 10 years, SNIC has been engaged in a national storage service for scientific data, called Swestore. While Swestore has grown into a petascale system, supporting many researchers in different fields of science, many new users have requested additional methods of access and user authentication as well as metadata management. The current Swestore service is based on dCache with GSI authentication. This technology was brought to Sweden alongside the

Nordic Tier-1 Storage Infrastructure for WLCG (Worldwide LHC Computing Grid).

With dCache Swestore is able to provide users with a reliable distributed storage service for their scientific datasets. However, for certain more advanced use cases such as (automated) metadata management, other solutions such as iRODS would be more suitable. With the deployment of iRODS, to complement existing Swestore service, these functional limitations of the current Swestore offering could be overcome. This was the motivation for the SNIC iRODS project, which is now in the process of being deployed into production. Within the project, spearheaded by Stockholm KTH PDC and Linköping NSC, we were able to build a scalable and distributed storage system with customisable features.

We developed methods for geo-replication of and failover of the iCAT database as well as methods for for deploying distributed storage resources within Sweden with iRODS. We have integrated the iRODS-based storage environment into the SNIC services side-by-side with the existing SNIC Swestore storage service. We also made some contributions to the iRODS ecosystem, namely: a novel tape interface for iRODS written against the IBM Spectrum Protect (TSM) API; s syslog forwarder which can be used with iRODS logs; an alternate PAM authentication interface with debugging features; and an automated provisioning environment to build iRODS grids with optional provisioning of (disposable) virtual machines for testing.

REDUNDANT GEO-REPLICATED ICAT

For resilience our resource hierarchies are configured for replication of the data and to make the grid redundant, the iCAT server is also (geo) replicated. The iCAT servers are running local PostgreSQL database servers, and using streaming replication to a hot standby which also has a warm standby iCAT server. The entire failover process is scripted, albeit it is manual on intent, since distributed systems are complicated and in this case we thought it is better for the system administrator to be in control, to be able to verify every recovery step when a disaster should happen.

For the iCAT server / PostgreSQL database we use a set of small scripts to initiate the failover. To switch over to the standby servers `dnsupdate` is used to change the IP addresses of the respective machines. We have a subdomain delegated for the iRODS services and we update the addresses in this zone. The zone has a slave in Stockholm to take over. The secondary PostgreSQL server can also be used for read-only queries such as accounting or detailed usage statistics.

In case of failure we trigger the PostgreSQL failover to promote the secondary to primary and we start the standby iCAT server. A monitoring script sets the unavailable resources offline so they will be served from the available replica resources. Finally we update the DNS entries to refer the takeover hosts.

MODEL FOR STORAGE RESOURCE DEPLOYMENT

Because of the distributed nature of SNIC services being coordinated over several HPC centers, SNIC storage services are also deployed with a distributed model. Currently the Swestore service is being operated jointly by several centers with NSC and HPC2N providing the core services and several other centers run dCache storage pools.

With the deployment of iRODS, we follow the same model with KTH PDC and Linköping NSC providing the core services and replicated storage pools in the start of operations. Later more centers can be included with their prospective storage pools. The main principle is to provide high availability of services with geo replication.

We currently rely on iRODS built in replication (coordinating resource) to ensure dual copies of data objects between the two storage pools. This provides an easy way to ensure the successful replication of data. However, this method comes with a drawback, the performance penalty of synchronous replication. Since iRODS handles replication in sequence instead of in parallel, receiving the data object on one of the replicated resources first and then replicating the data object while the client waits for completion of the put operation. With a very fast network connection between the resource servers this affects the overall throughput less, but in our case it is noticeable. Hopefully the upcoming 100 Gbit/s upgrade of both the KTH PDC network backbone and SUNET connectivity would mitigate this.

Additionally, we have considered using asynchronous replication by implementing the replication as iRODS rules with delayed execution via the rule exec queue. This method would increase the overall throughput of the system with a tradeoff in resiliency. Finally, data locality remains an issue we need to solve. To maximise throughput and minimize network traffic over SUNET, the clients should push the first replica of the data always into the resource which is in the same subnet, and then replicate over SUNET.

In addition to the resources available in the SNIC zone, via federation against EUDAT and KTH PDC iRODS zones, we enable the use of remote zone resources for distinct users granted access. For example a KTH researcher might want to have copies in the SNIC zone and PDC zone as well, since the PDC iRODS resources are in the process of being made visible to the KTH PDC compute clusters via local InfiniBand access. This will be elaborated on in the last 2 sections of this paper.

TAPE LIBRARY ACCESS WITH TSM INTERFACE

Tape mass storage access is a frequent requirement for large scale storage systems, and is even more vital for an archival system. We have long traditions of using tape libraries at both PDC and NSC. The tape libraries are used for backups and archives via the use of IBM TSM (currently called Spectrum Protect, but referred as TSM in this paper).

Since iRODS provides the Universal Mass Storage Interface driver to define a resource accessed via external shell scripts (or executables), we developed a simple utility which can be used with the MSS driver to provide tape storage access using the TSM API. It is implemented in C and it compiles into one executable. It requires a TSM client to be installed and configured, with the TSM API packages also installed and TSM credentials in place.

It implements the MSS operations, which are syncToArch, stageToCache, mkdir, chmod, rm, mv and stat. There are additional operations for housekeeping. It is prepared to handle a list of files to access them in the optimal order, however unfortunately iRODS issues the access requests only one-by-one. To create a tape resource we need a resource hierarchy with a cache resource in front. Then we simply define an MSS resource with the `dsmarc` executable as the driver in the context string of the resource.

The sources are available at `https://github.com/KTH-PDC/irods-dsmarc`.

PAM AUTHENTICATION

An alternate implementation of the iRODS `PamAuthCheck` module was also developed at our project, with more robust error handling and debugging features. PAM authentication provides extreme flexibility which makes the external PAM authentication module is a very useful feature in iRODS. But, for the very same reason PAM authentication can also become rather complicated. To facilitate troubleshooting in the PAM configuration, more verbose debugging and logging functionalities have been added in our implementation.

An additional feature has been implemented to enable per user choice of PAM service files for authentication. With an extra config file instead of the default `irods` PAM service, a different PAM service can be assigned to the user. This provides more choices for the authentication, i.e. one user can use traditional password authentication, while an another a hardware token and so on.

The source distribution is very small and simple, only one C source file and a makefile. With some additional work further enhancement could be possible to provide LDAP integration, so that extra LDAP group membership could decide which PAM service a given user should be associated with.

The sources are available at `https://github.com/KTH-PDC/irods-pamauth`.

FORWARDING OF IRODS LOGS

In a similar fashion to the previous utilities, a simple utility has been developed which watches a set of log files and forwards the new log entries to `syslog` as they are being appended by the application. The utility is implemented in C and it runs as a daemon in the background. The messages are sent to the local `syslog` daemon which will process and forward them if configured to do so.

The sources are available at `https://github.com/KTH-PDC/irods-logforw`.

AUTOMATION

We are consolidating our configurations in Git repositories, currently hosted in GitHub, in our `KTH-PDC` namespace. To later enable continuous integration and for us to be able to test and verify our configurations, we developed an Ansible environment for deploying entire iRODS grids with a built-in replicated iCAT and all services deployed. The package is called `irods-provisioner` and it contains a set of idempotent Ansible *roles* for deploying a certain service or function. Together these form a fully functional iRODS grid.

The deployment of a fully configured iRODS grid into a cluster of virtual machines in VirtualBox using Vagrant can be done as follows with `irods-provisioner`.

```
$ git clone -b 4-1-stable https://github.com/KTH-PDC/irods-provisioner.git
$ cd irods-provisioner
$ vagrant up
$ ANSIBLE_HOST_KEY_CHECKING=False ansible-playbook -b -i hosts-test irods-cluster.yml
```

The sources are available at `https://github.com/KTH-PDC/irods-provisioner`.

INTEGRATION INTO SNIC SERVICES

To integrate the new iRODS storage into the existing SNIC Swestore services, we integrated SUPR (SNIC User and Project Repository) into iRODS and FreeIPA, which we use as an Identity Management (IdM) solution.

SUPR and iRODS Integration

SUPR is an online user management and project application/approval system for SNIC. Synchronising user and project data from SUPR to iRODS enables us to have a master source for users and PIs to manage their data and projects both from iRODS and other systems. Prior to this synchronisation, the project request and approval was carried out through support tickets, which can be tough to track for changes.

The SUPR-iRODS synchronisation script uses the Python iRODS Client API supported by the iRODS Consortium. It connects to SUPR periodically and checks for new iRODS projects and users, pulls the user data, creates the user id associated with the user name and creates users, groups and collections for the user(s) in the SNIC iRODS.

SUPR and FreeIPA Integration

FreeIPA is a identity management system (IdM). We use it for SNIC iRODS (and later most likely the entire Swestore) centralised user and password management, to enable the users to manage their user accounts and passwords for both dCache and iRODS. We were able to integrate SUPR and FreeIPA so that the iRODS users from SUPR are synchronised to FreeIPA. The users can then log in to FreeIPA and can set the password for dCache or iRODS.

FEDERATION

The SNIC iRODS grid is federated with the Swedish EUDAT zone (also operated by KTH PDC) as well as the KTH PDC local iRODS zone, for enabling certain users the access of additional resources. This way we are able to deploy an iRODS path from local parallel filesystems at the HPC resources via national resources to European resources.

HPC Integration

At KTH PDC we are deploying a local iRODS grid for HPC users. The most important objective of this iRODS deployment thus is performance. The storage resources at the `pdc.kth.se` iRODS grid are to be accessible via the local InfiniBand fabric via IPoIB. We are developing proof-of-concept models for deploying high performance storage for iRODS resource servers in a scalable and cost-effective fashion. We are aiming at high performance transfers in and out of our 5 PB Lustre filesystem to offload the cluster filesystem storage.

We developed an effective proof-of-concept solution with InfiniBand SRP and ZFS. We are able to access multiple ZFS pools via multipathing over redundant links and separate IB fabrics. With this scalable approach we are able to deploy more JBOD storage to the fabric(s) and by increasing the number of ZFS/iRODS servers more throughput will be available.

Next step in our testing will to introduce the ZFS resource servers to our 100 Gbps EDR InfiniBand fabric at our pre/postprocessing cluster. We will be testing IPoIB performance with iRODS as well as bare IP over 100 Gbps Ethernet parallel transfer performance. So far we were able to reach at maximum $\approx$ 3,300 MB/s iRODS parallel transfer throughput. This is however achieved with incomplete TCP and hardware tuning (iperf $\approx$ 42 Gbps max).

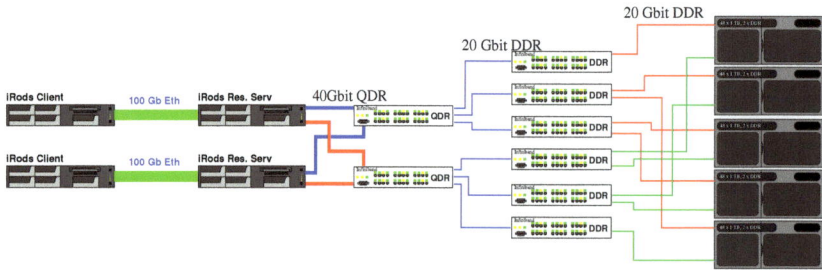

Figure 2. KTH PDC iRODS over 100 Gbps Ethernet testing environment with an InfiniBand SAN

A national approach for storage scale-out scenarios based on iRODS

Christine Staiger
SURFsara
Science Park 140,
Amsterdam, The Netherlands
christine.staiger@surfsara.nl

Ton Smeele
Utrecht University
ITS/RDM
Heidelberglaan 8,
Utrecht, The Netherlands
a.p.m.smeele@uu.nl

Rob van Schip
Utrecht University
ITS/RDM
Heidelberglaan 8,
Utrecht, The Netherlands
r.j.a.vanschip@uu.nl

ABSTRACT

The Dutch Universities and associated Medical Centers are developing research data management environments built on iRODS to support their scientists. The underlying storage is currently primarily located on the premises and under the control of said institutes. However, some local storage systems offer too little capacity. Moreover, there is a need for a variety of storage systems to offer efficient and cost effective data storage solutions that may differ per use case. Because requirements towards the storage backend between single research institutes overlap, a national approach can add significant value. We present a proof of concept study how such a scenario can be supported using iRODS. In our use case scenario SURFsara, the national high-performance compute (HPC) and data centre, provides storage resources connecting local data to European infrastructures such as EUDAT, EGI and PRACE. We highlight the infrastructural aspects and which data policies can be supported. The scenarios are substantiated with performance tests executed with the underlying transfer protocol to the different storage systems.

Keywords

iRODS, Storage scale-out, Infrastructure, Policies, Research data management (RDM) platforms.

INTRODUCTION

The Dutch Universities are developing data management platforms to assist researchers to safely store and collaborate on data during and after their research, to account for data generated and processed in a research project and to facilitate reuse of such data. The way how data needs to be treated and stored, i.e. the data policies, can vary per university, per faculty and per research project.

We base our study on data management platforms built on iRODS drawing on the following advantages:
Data policies can be system-enforced, targeted to specific data types and use cases and maintained efficiently. Moreover, iRODS allows to integrate heterogeneous storage solutions accounting for different requirements and lowering costs for storage by combining cheap and expensive storage media. Management, distribution and migration of data files across locations and vendor storage solutions is performed transparent to the user and automated through and directed by applicable data policy rules.

Additional cost reduction can potentially be achieved through adopting a cloud storage delivery model that serves storage for several projects, institutes and universities lowering overhead costs as well as employing expertise on different storage systems from a dedicated third party rather than fostering and maintaining expertise and hardware at the single Universities and institutes.

Research projects are dealing with sensitive data. Those data are subject to strict legal regulations, e.g. such data must be managed by University staff and may not be transferred across national or European Economic Area (EEA)

iRODS UGM 2017 June 13-15, 2017, Utrecht, Netherlands
[Author retains copyright. Copyright © 2017 Christine Staiger, SURFsara Amsterdam, The Netherlands; Ton Smeele and Rob van Schip, Utrecht University, The Netherlands.]

borders. The data management platforms need to support such strict data policies which are very hard to put into practice when combining those platforms with storage from commercial storage providers.

As the collaborative ICT organisation for Dutch education and research, SURF [1] is part of the Dutch research landscape and part of the European research infrastructures such as PRACE [2], EGI [3] and EUDAT [4]. Thus, SURFsara [5], as part of SURF, can serve as a trusted storage provider and is well-positioned to support the scenario above. We will investigate the opportunity to provide a cloud storage solution as a service managed by SURFsara that integrates with each university's iRODS data management platform.

Such a cloud storage solution needs to support a replication and a storage scale-out scenario. In a replication scenario universities outsource secondary copies of data to storage provided by SURFsara to serve as fall back copies for disaster recovery. In a scale-out scenario, however, universities store active data on such a storage system, i.e. scientists work directly with these data. In fact, in both cases the same infrastructure can be used. In our investigation we focus on the scale-out scenario, since this is the most demanding scenario with respect to performance requirements.

We present a proof-of-concept study that can support the above-mentioned scenarios. We provide the technical setup for both scenarios and we test in particular the scale-out with respect to performance and user experience and whether the local data policy [6] requirements can be met.

Note, out of scope of our study are use cases that benefit greatly from using storage directly attached to a workstation because they have been designed to take advantage of low latency disk read/write operations.

USE CASES

The data management platforms and thus the underlying infrastructure are built for scientists to maintain their data during and after the research process. We will discuss and test the following use cases where performance plays a paramount role:

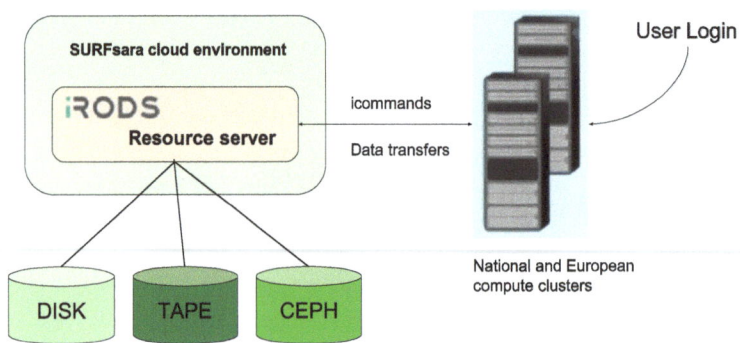

Figure 1. Usage of data managed by iRODS from compute systems.

- **Users mount iRODS to their workstations to up and download data and to work on the data directly.** This is accomplished by Davrods [7] and allows users to drag and drop data between the iRODS file system and their local file system. Data can be stored on local storage or on a scale-out resource server at a different site. For programmatic data transfers users employ the *icommands* to put and get data to and from iRODS.

- **Users manage data in iRODS and analyse data on HPC infrastructures (Figure 1).** To this end an HPC cluster at an HPC centre such as SURFsara is used. SURFsara hosts the national supercomputer [8] that is part of PRACE (Partnership for Advanced Computing in Europe Research Infrastructure) and the national compute cluster [9]. The use of a storage service close to the HPC infrastructure can improve transfer speed.

The HPC cluster needs to accommodate an iRODS client e.g. the icommands with which the user can move data between the iRODS instance and the HPC cluster.

For another use case we will discuss the technical setup:

- **Long-term archiving of data** can be accomplished by storing data on tape and labeling it for later reference. Data will be migrated to cheap, high-latency media as tape. Here a replica or copy is created on the iRODS resource server.

PROOF OF CONCEPT ARCHITECTURE

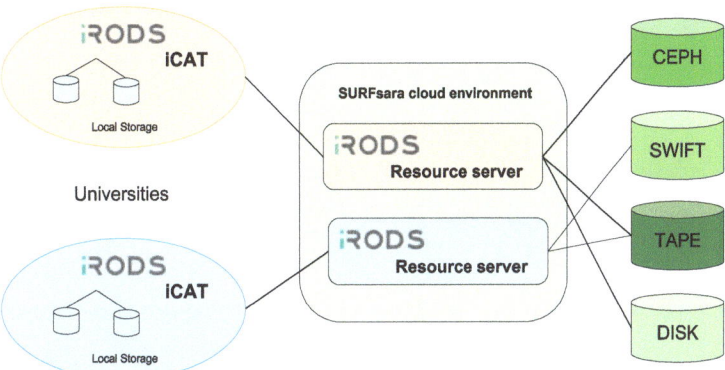

Figure 2. High-level overview of the infrastructure. Universities get access to storage infrastructure via iRODS resource servers which are attached to the University's iRODS zone. The resource servers are hosted on virtual machines running on a SURFsara cloud environment.

To facilitate the access to the national storage systems and integrate them with the Universities' iRODS platforms we deploy iRODS resource servers on the SURFsara HPC cloud (see Figure 2). Storage can be attached to these resource servers as first order resource or as compound resource depending on the backend storage system. Universities get access to storage infrastructure via iRODS resource servers which are attached to the University's iRODS zone. This guarantees that all data is subject to the Universities' data policies although located at a third-party storage provider. In the following paragraph we will describe how scientists and data managers can make use of the underlying infrastructure.

Storage system implementations

In the following sections we will describe the technical setup for several storage systems. Our tests are based on an implementation using a CEPH storage system. In our setup all resource servers are run on SURFsara's HPC cloud environment based on OpenNebula. From there the connection to other storage media is made.

Storage systems that support POSIX compliant random access file operations can be attached directly as a *unix-filesystem* type resource. Other storage systems such as object stores and tape archival systems typically need to be configured as a compound resource. The compound resource adds a POSIX compliant cache resource in front of an archive resource. Exchange of data objects between cache resource and the actual storage system is executed through vendor specific drivers that perform data transfer protocol translations.

CEPH

The HPC cloud infrastructure uses a CEPH cluster to support virtual machines with extra storage. CEPH partitions are attached as an extra file system to the virtual machines directly. Such storage can be integrated in iRODS as iRODS *unixfilesystem* resource (see Figure 3, left side). Most other resources like archive and SWIFT will be accessed via a compound resource as we will see in the following Sections.

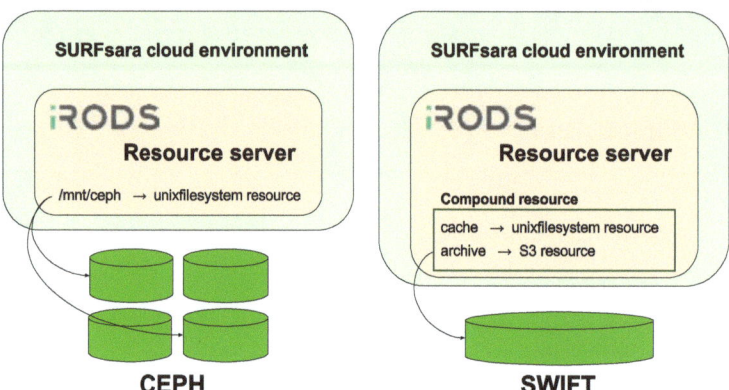

Figure 3. Storage scale-out by either as first order resource or via a compound resource. Left: By attaching file systems to the resource server these file systems can be used by an iRODS *unixfilesystem* storage resource. Right: The connection to an OpenStack SWIFT cluster can be made via iRODS compound resources employing the iRODS S3 resource type as archive resource. Green: Storage and infrastructure managed by SURFsara; Orange: iRODS servers managed by the universities.

Openstack SWIFT

In contrast to CEPH storage, iRODS can only make use of OpenStack SWIFT or other S3 compatible storage types via a compound resource (see Figure 3, right side). The connection between the cache resource and the OpenStack SWIFT archival resource is made via the S3 plugin. The plugin uses a login on the OpenStack SWIFT cluster in form of an AWS key-pair, i.e. all data on this storage will be owned by this account no matter which iRODS user ingested the data into the resource.

Archiving data in a Tape library

To account for the need of cheap storage that supports long-term archiving of data we integrated the resource server with an storage environment based on tape. This environment can be accessed from iRODS via a compound resource as we saw with OpenStack SWIFT. However, in this case the iRODS distribution does not provide a native plugin to facilitate the communication between the cache resource and the archive resource. The communication between the cache and archive resource is defined by a universal MSS interface script that implements the functions *syncToArch*, *stageToCache*, *mkdir*, *chmod*, *rm*, *mv* and *stat*. Based on the general universal MSS interface [10] SURFsara provides such universal MSS interface scripts to connect to tape environments using either *gridFTP* [11] and *rsync* [12].

Federation as an alternative for storage system implementations

Opposed to the previous architecture where we extended the storage under one iRODS instance by directly attaching resources, one can also make use of iRODS federations to give access to the underlying storage infrastructures. Federations are not a solution for storage scale-out, yet federations support replication scenarios. We will briefly outline an example of such an architecture below.

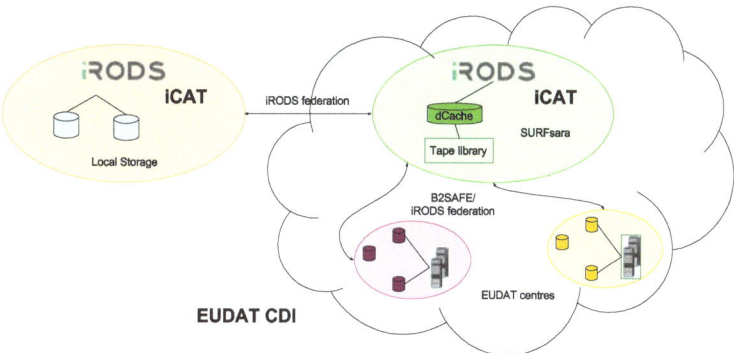

Figure 4. Archiving via a federation. SURFsara hosts an own iRODS instance which is coupled to the tape library via the dCache environment. The same iRODS instance is part of the EUDAT CDI and implements the B2SAFE service with which data can be replicated to other EUDAT centres.

SURFsara hosts an own iRODS instance. This instance is part of the EUDAT B2SAFE [13] network and part of EUDAT's collaborative data infrastructure (CDI). B2SAFE is EUDAT's service for safe data replication between EUDAT centres. The service integrates iRODS with persistent identifiers to keep track of data and its replicas. Hence, Universities can use the B2SAFE service to create persistent identifiers for the replicas for identification and citation; and use SURFsara's iRODS instance as an entry point to the EUDAT CDI as indicated in Figure 4.

RESULTS AND DISCUSSION

Evaluation of the implementation

In the evaluation of the proof of concept implementation we focused on the usability of the CEPH resource under the iRODS resource server.

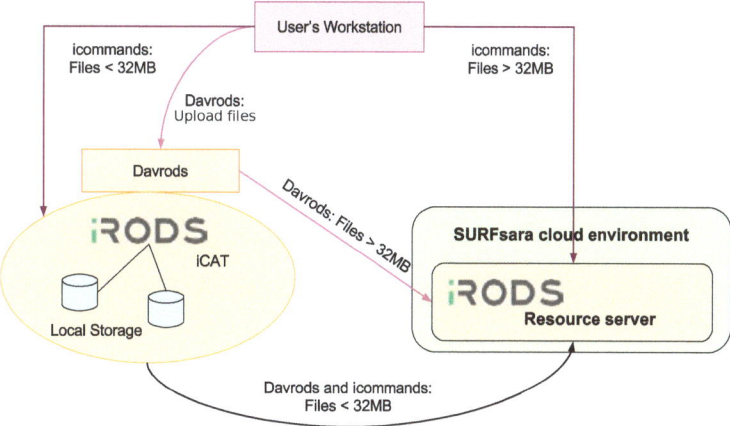

Figure 5. Testing scenario for interacting with the iRODS resource server from a user's workstation. User connects with Davrods to the iCAT enabled server (light purple arrow). For file transfers less than 32 MB the iCAT server acts as a hub in between Davrods and the resource server (black arrow). Transfer of larger files is performed via peer-to-peer connection between Davrods and the resource server. As an alternative to Davrods, *icommands* can be deployed on the user's Linux workstation (dark purple arrows). In that case larger files are transferred peer-to-peer between workstation and resource server.

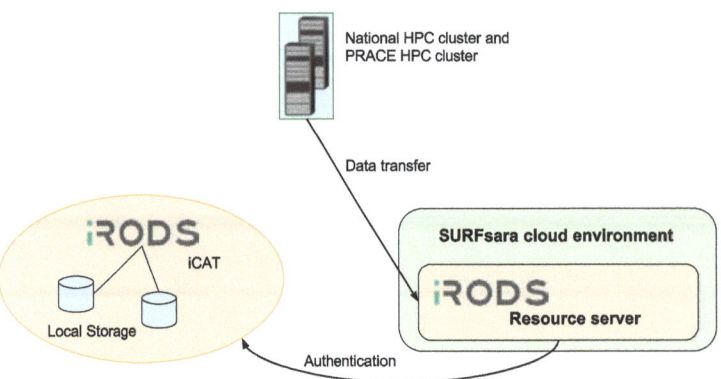

Figure 6. Testing the up and download of files to the resource server with the *icommands* from the HPC clusters. The user connects directly to the iRODS resource server which will in turn connect to the iCAT database to authenticate the user. The data is directly up and down loaded to and from the iRODS resource server.

To gain insight into the overall user experience, we tested typical workflows from a local workstation using the CEPH resource (see Figure 5). Furthermore, we measured up and download speeds of file transfers to and from the resource server

- from a local workstation with the *icommands* (see Figure 5)
- from the national and European HPC clusters employing the *icommands* (see Figure 6)

User experience tests

To test end user experience we executed several office application workflows using two different client workstations and three different storage locations. Our client environments include both Windows7 and Linux client workstations to identify potential impact of client operating system on the user experience. Our storage locations are 1) workstation locally attached disk drive 2) a storage resource directly managed by and attached to the university's iRODS iCAT server and 3) the CEPH storage partition located at SURFsara in Amsterdam managed via a resource server. The iCAT server communicates with the resource server via the internet.

Upon each test our client workstation connects to the iRODS iCAT server via Davrods and mounts the iRODS home collection as a network drive. We worked with files stored locally on the workstation and compared this experience with working with files stored on the network drive. As for the network drive, we varied the default resource configuration of Davrods to select either the resource on the iCAT-enabled iRODS instance or the resource on the iRODS resource server.

We used the MS-Office suite on Windows 7 and LibreOffice suite on Linux to manipulate text documents and spreadsheets. We also used accessory tools such as an ascii-text editor and a web browser to browse ascii text, JPG and PDF files. The files varied in size between 15 kilobytes and 17 megabytes.

We found that open and save operations that access a file stored locally on the workstation are slightly faster than using similar operations to access a file stored on the iCAT server resource. In nearly all workflows the response times remained below a second and as such they are within acceptable user experience ranges. In odd cases the response time in the iCAT server resources configuration could amount to 2-3 seconds. Interestingly, the response time remains about the same when we change our configuration to access a file stored on the resource server.

The choice of the client workstation operating system did not influence the user experience in all major operations that we tested with one significant exception: it took upto 10 seconds to mount the network drive for the first time using the native MS Windows drivers (Windows Security). Alternatively when we opened a connection on the same workstation using Cyberduck drivers the operation completed within 3 seconds.

On the whole this experiment shows that the performance in all configurations will support the user in working with his/her data in an adequate way.

File transfer tests

We tested file transfers to and from the iRODS resource server using *icommands* in two settings: 1) from a Linux workstation and 2) from two HPC compute clusters. In the first setting the user connects to the iCAT-enabled iRODS server but stores data on a CEPH resource located at the iRODS resource server (Figure 5). In the second setting (Figure 6) the user connects from the HPC clusters directly to the iRODS resource server and stores data on the respective CEPH resource.

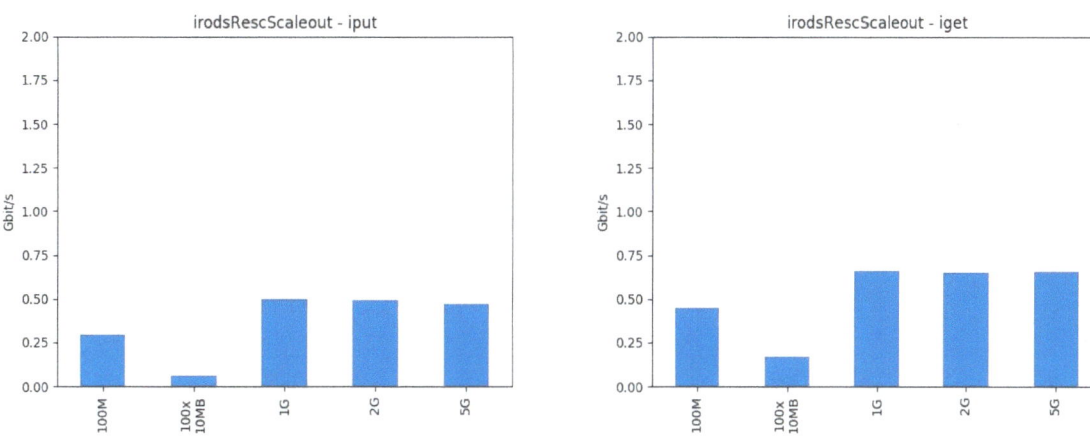

Figure 7. Mean performances of iput (left) and iget (right) from a user's Linux workstation to the iRODS resource server via the iCAT-enabled server.

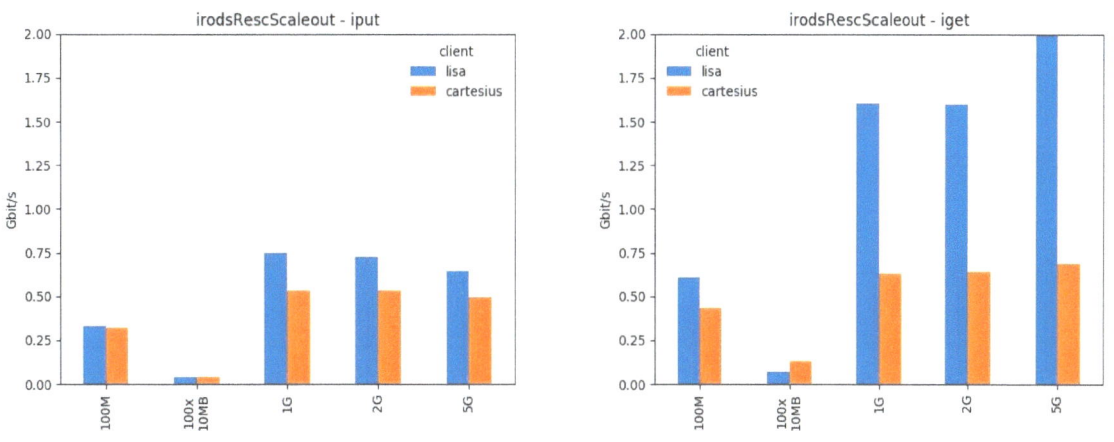

Figure 8. Mean performances of iput (left) and iget (right) from the national supercomputer (cartesius) and from the national compute cluster (lisa). The difference in transfer speeds can be explained by the different internet network connections on the two infrastructures.

Figure 7 and 8 show the performance of put and get operations on the iRODS resource server from the respective HPC clusters and the workstation. We tested the up and download of single files of size 100MB, 1GB, 2GB and 5GB and the transfer of a collection containing 100 files of size 10MB. The tests show that the data transfers to and from the CEPH resource work slightly faster when the client connects directly the resource server. Notably, up and downloading collections with a lot of small files (100x10MB) is much slower than downloading the same amount of data in a single file (1G). The low performance is due to setting up connections for each single file and look-ups and entry creation in the iCAT database. Hence the experiment shows that the iRODS resource server configuration efficiently supports use cases where larger files are transferred. Use cases that involve transfer of batches of smaller files will require additional measures to counter protocol overhead (e.g. bundle files prior to transfer).

Other findings

Impact on network configurations

Connectivity tests have shown that iRODS expects iCAT and resource servers in the data grid zone to be accessible using their fully qualified domain name (FQDN). Network configurations where resource servers are addressed via a proxy server such as a load balancer are not fully supported.

This limitation is a result of the iRODS parallel transfer protocol implementation which by default kicks in on transfers of files that exceed 32 MB in size. Suppose a client connects to server A. Now consider a scenario where the file needs to be transferred to or from a resource not managed by server A. In this case server A will use the ICAT database to locate server B that manages the resource. It opens a server-to-server connection to server B. Server B provides its own hostname (FQDN) and the TCP ports to be used for data transfer. Server A communicates this information to the client so that the client can open ports to server B. Subsequently the data flows directly between the client and server B. Note that in this scenario the client must be able to connect directly to server B using B's FQDN which could fail if server B is behind a proxy.

Impact of compound resources

In the previous sections we have shown that other complex storage systems can only be made available to iRODS via a compound resources. This has impact on the data workflows in such a storage system. When uploading data directly to a resource (e.g. a *unixfilesystem* resource) the user can be sure that the data is stored correctly after the transfer finished. In case of uploading data to a compound resource the user can only be sure that the data is stored safely on the cache resource and will eventually - depending on the system configuration - be moved to the archive resource. This poses two risks:

1. If the connection to the archive resource does not work as expected, the cache resource is filled up and no data is further transferred, clogging the system for other users and not keeping the data safe.

2. The user himself has to either rely on the system configuration to delete the replica located on the cache resource as soon as possible or he has to do that himself, which in turn requires users to be familiar with the underlying infrastructure and the command line tool options for the *icommands*.

The usage of the tape environment is possible in two ways: either directly using a compound resource or indirectly using a federated zone that employs the compound resource. Attaching the tape environment as a compound resource to iRODS allows Universities to integrate this resource seamlessly into their environment and manage access with their data policies. Alternatively, federations allow for complex configurations across administrative domains.

CONCLUSIONS AND FUTURE WORK

We demonstrated that a national cloud storage service can be used as a seamless extension of iRODS-based data management platforms hosted by the Dutch Universities and research institutes. Read and write performances remain within acceptable user experience ranges except that transferring batches of small files is relatively slow. Deployment of a national cloud storage service in a scale-out scenario requires that the Universities' iRODS servers are directly accessible from the internet.

In our work we did not test performances in a real-life setting, i.e. many users, large amounts of files. In the future we plan to explore these scalability aspects. Moreover, specific service configurations need to be tested e.g. what are the performance characteristics and limitations when using a compound resource. We will also investigate in which ways the cloud storage service model can be complemented by other service models based on zone federations rather than zone extension.

REFERENCES

[1] SURF. Collaborative organisation for ICT in dutch education and research. [Online]. Available: https://www.surf.nl/en

[2] PRACE research infrastructure. [Online]. Available: http://www.prace-ri.eu/

[3] EGI. Advanced computing for research. [Online]. Available: https://www.egi.eu/about/

[4] EUDAT. The EUDAT collaborative data infrastructure. [Online]. Available: https://www.eudat.eu/eudat-cdi

[5] SURFsara. High performance computing & data infrastructure for science and industry. [Online]. Available: https://www.surf.nl/en/about-surf/subsidiaries/surfsara/

[6] YODA. University Utrecht iRODS rule set. [Online]. Available: https://github.com/UtrechtUniversity/irods-ruleset-uu

[7] T. Smeele and C. Smeele, "Davrods, an apache webdav interface to iRODS,âĂŹâĂŹ in *Proceedings iRODS User group meeting 2016*. https://irods.org/uploads/2016/12/irods_ugm2016_proceedings.pdf, 2016, p. pp 41.

[8] Cartesius: The Dutch supercomputer. [Online]. Available: https://userinfo.surfsara.nl/systems/cartesius

[9] The Dutch national compute cluster: The lisa system. [Online]. Available: https://userinfo.surfsara.nl/systems/lisa

[10] J.-Y. Nief. Universal mss interface. [Online]. Available: https://github.com/cookie33/irods-compound-resource/blob/master/scripts/univMSSInterface_generic.sh

[11] R. Verkerk. iRODS composable compound resource description at SURFsara. [Online]. Available: https://github.com/cookie33/irods-compound-resource

[12] C. Staiger. iRODS compound resource training. [Online]. Available: https://github.com/EUDAT-Training/B2SAFE-B2STAGE-Training/tree/master/ExampleTrainings/iRODS-SysAdmin-Training

[13] EUDAT. B2SAFE. [Online]. Available: https://eudat.eu/services/userdoc/b2safe

Davrods enhancements as part of the Grassroots Infrastructure

Simon Tyrrell
The Earlham Institute
Norwich Research Park,
Norwich, NR4 7UZ, UK
simon.tyrrell@earlham.ac.uk

Xingdong Bian
The Earlham Institute
Norwich Research Park,
Norwich, NR4 7UZ, UK
xingdong.bian
@earlham.ac.uk

Robert P. Davey
The Earlham Institute
Norwich Research Park,
Norwich, NR4 7UZ, UK
robert.davey@earlham.ac.uk

ABSTRACT

This paper explains the enhancements we have made to the Davrods Apache module to expose a range of iRODS functionality that was previously unavailable, and configuration improvements to allow the default interface to be made more user-friendly. We have made changes to Davrods so that iRODS can be used as the storage mechanism for public facing websites without the need for users to be authenticated, making it easier to produce web-based data repository interfaces. We provide code to expose iRODS metadata as cross-referencing links between data objects and collections. We also describe a REST API that has been added for metadata functionality within iRODS to facilitate metadata manipulation by end users with the supplied client-side code from within their web browsers or from other web services.

Keywords

iRODS, Apache httpd web server, infrastructure, web client, RESTful web service

Introduction

The Grassroots[1] Infrastructure project aims to create an easily-deployable suite of computing middleware tools to help users and developers gain access to scientific data infrastructure that can easily be interconnected.

With the data-generative approaches that are increasingly common in modern life science research, it is vital that the data and metadata produced by these efforts can be shared and reused. The Grassroots Infrastructure project wraps up industry-standard software tools along with our own custom open-source software tools to give a consistent API that can be federated with others in terms of both data and services. This means institutions and groups can deploy a simple lightweight software suite, locally or as a virtual machine, to expose institutional data, connect up any existing data services, and federate their instance of Grassroots with other remote instances.

One of the major aims of the Grassroots Infrastructure is to allow users to share their wheat data, although it is by no means organism-specific, as easily and seamlessly as possible. For data storage, we use the iRODS[2] data grid system that gives users access to potentially differing file systems and data resources through a single data abstraction layer. Users are able to carry out typical filesystem actions as normal, such as creating files and directories and maintaining permissions, but with extra features such as distributed storage viewable across different institutions and the ability to add extensive metadata to files and directories.

iRODS ships with command-line clients to provide access to data storage managed by the platform, and many Application Programming Interfaces (APIs) exist in a variety of languages to support programmatic development such as PyRods[3] and Jargon[4]. An Apache httpd[5] module based on the WebDAV protocol, Davrods[6], exists to allow access to iRODS data stores using standard WebDAV commands. This project supports much of the basic functionality of a web-based data dissemination stack, but there were a set of missing features of Davrods that we have developed that can improve data searching, as well as the general user experience.

iRODS UGM 2017 June 13-15, 2017, Utrecht, Netherlands
Copyright (c) 2017, The Earlham Institute

Themed listings

The standard web pages produced by Davrods resemble the basic directory listings produced by Apache. Whilst simple and effective, the functionality of these pages is lacking, and there is little configuration ability to make their look and feel customisable. Therefore, using the concepts from the autoindex module[7], we have developed a mechanism to insert themes into Davrods listing pages using typical HTML and Cascading Style Sheets (CSS) elements. Three points in the web page were identified to allow the insertion of custom HTML chunks: the head section of the web page and the sections before and after the iRODS directory listings. These HTML chunks can either be set as strings in the Apache configuration file or point to separate files on disk to allow for easy modification. Any changes made to the files are instantly available for all subsequent requests whereas any changes to the string-based configuration require a restart of the Apache server. Additionally, the columns of the listings table have been marked up with consistent CSS classes to allow for easy customisation by server administrators when developing CSS for use on their own project or institution pages. For each of its generated pages, Davrods includes a link to the parent collection as a form of breadcrumbs. However if the iRODS instance has a multi-level hierarchy in its iRODS zone this can become unwieldy as a user can only travel up one level in the hierarchy at a time. We replaced this single level breadcrumb with a navigation element containing the full breadcrumb chain for the set of parent collections up to the root collection, thus resembling the navigation concept commonly used on many websites.

The autoindex module also includes the ability to specify default icons for various arbitrary data types, typically denoted by file extension. We have developed this within Davrods so that custom icons can be used for the iRODS data objects and collections. This can be further refined and default icons can be specified for data objects with unknown file types. For example, to use the icon stored within the image file at **/davrods_files/images/picture** for PNG, GIF and JPEG images, the directive would be:

```
DavRodsAddIcon /davrods_files/images/picture .png .gif .jpg .jpeg
```

An example screenshot of various parts of the theming functionality is shown in figure 1.

Public access

The iRODS data workflow is based around the concept that users log in to see the data that they have permissions to access. Often for public websites serving open access data, there is a desire to give full read access to browse a list of files and directories without the need for dedicated login credentials. We have added the ability to have a default user to be specified within the Apache server configuration, that would be used to log into the iRODS system without the need for any user intervention. For example, to specify that the iRODS user called public_user, with a password of anonymous, is used, the following configuration would be used:

```
AuthType None
Require all granted
DavRodsDefaultUsername public_user
DavRodsDefaultPassword anonymous
```

In effect, this creates a public-facing website using the data stored in an iRODS storage system that is indistinguishable from a site generated from a regular directory on the local filesystem.

Metadata

One of the major benefits of iRODS is its ability to add metadata as attribute-value pairs to any data objects or collections stored within it. Previously with Davrods, any metadata held in an iRODS instance was not exposed. We therefore provide Davrods functionality to make metadata both viewable and editable, as well as a REST API interface to allow programmatic interrogation and modification of iRODS metadata. For giving read-only access to the metadata, the configuration directive **DavRodsHTMLMetadata** is used and it takes one of the following values:

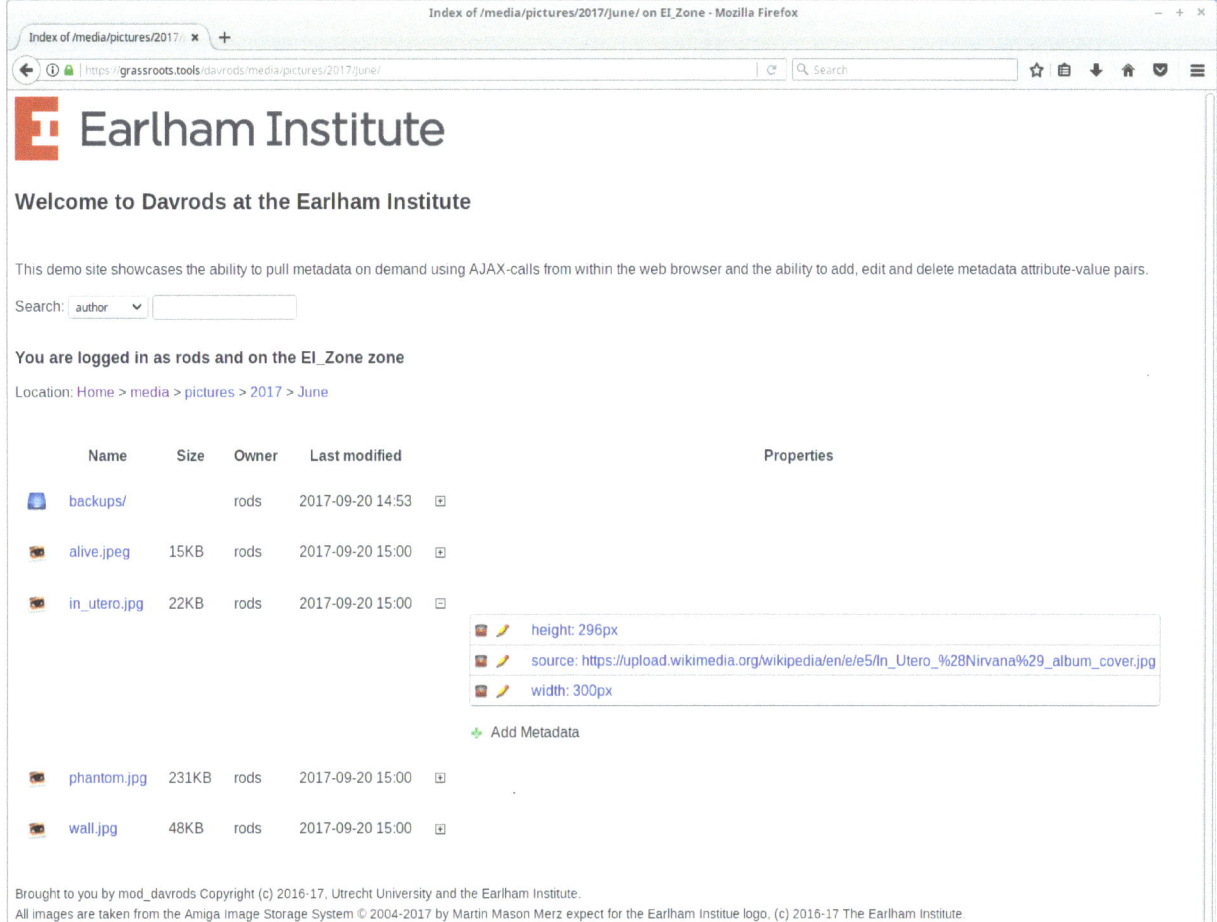

Figure 1. The themed interface showing the breadcrumbs, icons and HTML sections that can be configured

- **off**: The metadata display is disabled.

- **full**: All of the metadata is displayed for each data object and collection.

- **on_demand**: None of the metadata is initially included with the HTML pages sent by Davrods. Instead it can be delivered upon demand and inserted into the web page via AJAX requests from when the user clicks on the appropriate link associated with a given data object or collection.

Searching and linking

Each of the metadata attribute-value pairs are exposed as hyperlinks giving users a straightforward and standard method to find data objects or collections also containing the same pair. Additionally there is a general search mechanism provided to search across the entire metadata collection and this is available as a form within each page that Davrods delivers.

REST API

As well as adding a read-only view of the metadata, the ability to add, edit or delete metadata from within in a web browser for users with the appropriate permissions is also provided. For this feature to be active, the following

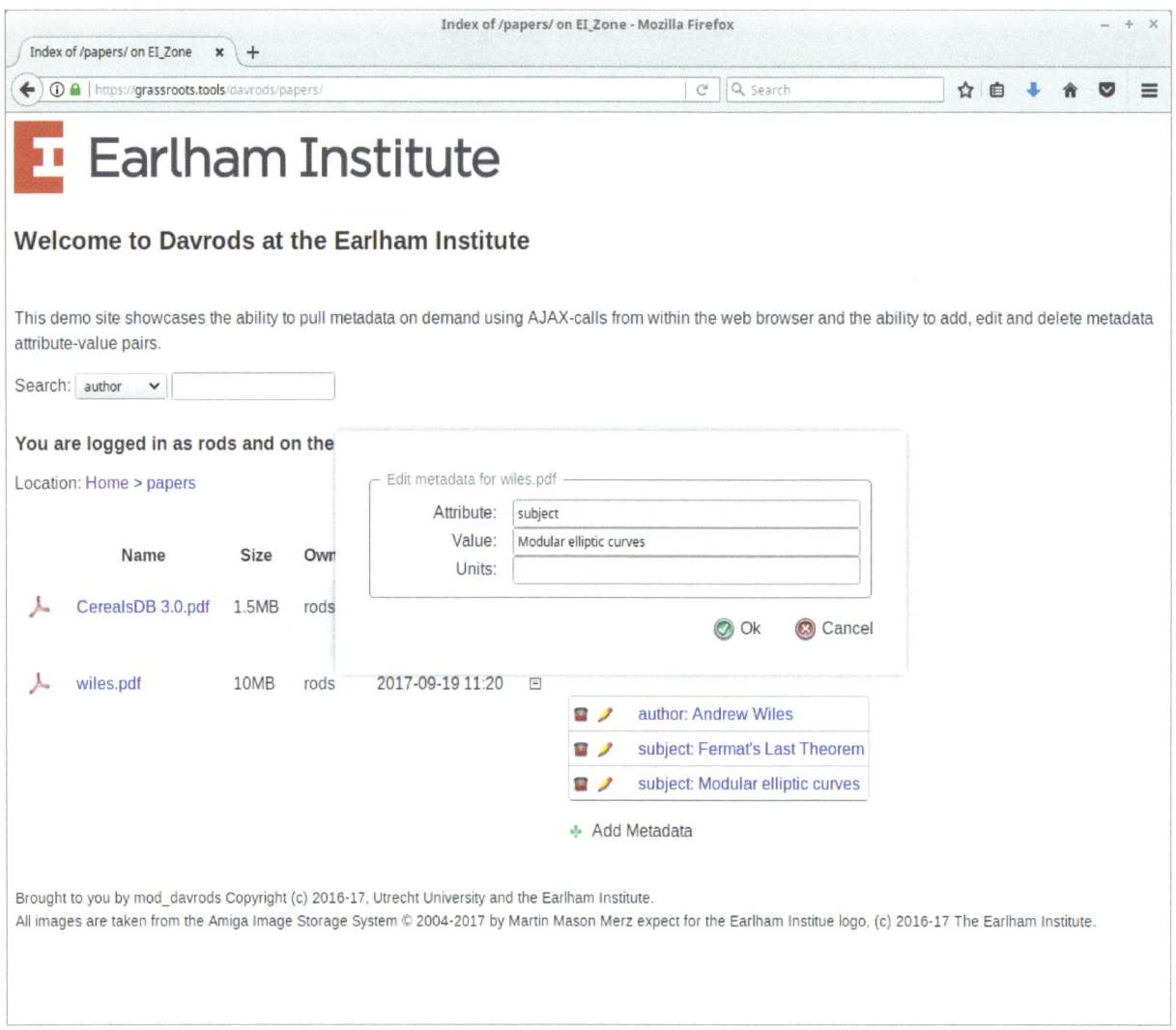

Figure 2. The metadata editor

configuration directive needs to be set:

```
DavRodsHTMLMetadataEditable true
```

The manipulation of metadata is facilitated by a REpresentational State Transfer (REST) API for querying and manipulating the iRODS metadata. The base URL for this API is at `/api/metadata` although the prefix can be changed by altering the `DavRodsAPIPath` parameter in the Apache configuration. The REST API contains the following functions:

- **get**: Retrieve all of the associated metadata for an iRODS item. It takes a single parameter, *id*, which is the iRODS id of a data object or collection to be queried. For example, to get the metadata for a data object with the id of 1.10021, the URL to call would be:

    ```
    api/metadata/get?id=1.10021
    ```

- **search**: Retrieve a list of all data objects and collections that have a given metadata attribute-value pair. It takes two parameters: *key*, the attribute to search for, and *value* which specifies the metadata value. There is a third optional parameter, *units*, for specifying the units that the metadata attribute-value pair must also have. So, to search for all of the data objects and collections that have an attribute called volume with a value of 11:

    ```
    api/metadata/search?key=volume&value=11
    ```

- **add**: Add a metadata attribute-value pair to a data object or collection. It takes three parameters: *id*, the iRODS id of a data object or collection, *key*, the attribute to add, and *value* which specifies the metadata value to be added. As with the `search` call listed above, there is a fourth optional parameter, *units*, for specifying the units that the metadata attribute-value pair will have. So, to add an attribute called volume with a value of 11 to a data object with the id of 1.10021:

    ```
    api/metadata/add?id=1.10021&key=volume&value=11
    ```

- **edit**: Edit a metadata attribute-value pair for a data object of collection and replacing one or more of its attributes, values, or units. It takes the following required parameters: *id*, the iRODS id of a data object or collection, *key*, the attribute to edit, and *value* which specifies the metadata value to edit. There is an optional parameter, *units* for specifying the units that the metadata attribute-value pair must also have to match. There must also be one or more of the following parameters to specify how the metadata will be altered: *new_key*, which is for specifying the new name for the attribute, *new_value*, for specifying the new metadata value and *new_units* for specifying the units that the metadata attribute-value pair will now have. So, to edit an attribute called *volume* with a value of 11 and units of decibels, for a data object with the id of 1.10021 and give it a new value of 8 and units of litres:

    ```
    api/metadata/edit?id=1.10021&key=volume&value=11&units=decibels&new_value=8&new_units=litres
    ```

- **delete**: Delete a metadata attribute-value pair from a data object of collection. It takes three parameters: *id*, which is the iRODS id of the data object or collection to delete the metadata from, *key*, which is the attribute to delete for and, *value*, which specifies the metadata value to delete. As before, there is an optional parameter, *units* for specifying the units that the metadata attribute-value pair must also have to be deleted. So to delete an attribute called volume with a value of 11 and units of decibels from a data object with the id of 1.10021:

    ```
    api/metadata/delete?id=1.10021&key=volume&value=11&units=decibels
    ```

We have included a set of JavaScript functions to allow a davrods administrator to easily give a user to access each of these API functions from within the browser. An example screenshot of the editor is shown in figure 2.

Future Work

Currently the REST API functions all return HTML fragments. However, in future, we would like to develop the possibility of specifying other datatypes such as JSON fragments to allow for integration and automation with other consuming web services.

Acknowledgements

The Grassroots project is strategically funded through the BBSRC cross-institute Designing Future Wheat programme grant, BB/P016855/1, and aims to develop a lightweight reusable set of open source software tools to allow researchers to share and federate life science datasets.

Availability

The source code is available at https://github.com/billyfish/davrods.

REFERENCES

[1] Grassroots Infrastructure, `https://grassroots.tools`, Visited last on 06.06.2017
[2] Hao Xu, Terrell Russell, Jason Coposky, Arcot Rajasekar, Reagan Moore, Antoine de Torcy, Michael Wan, Wayne Shroeder, Sheau-Yen Chen: iRODS Primer 2: Integrated Rule-Oriented Data System. Synthesis Lectures on Information Concepts, Retrieval, and Services, Morgan Claypool (2017)
[3] Python iRODS Client (PRC) `https://github.com/irods/python-irodsclient`, Visited last on 06.09.2017
[4] Jargon, `https://github.com/DICE-UNC/jargon`, Visited last on 06.09.2017
[5] The Apache HTTP Server Project, `http://httpd.apache.org/`, Visited last on 06.06.2017
[6] Ton Smeele, Chris Smeele: Davrods, An Apache WebDAV interface to iRODS. iRODS UGM 2016 proceedings, pp. 41-47 (2016)
[7] Apache Module mod_autoindex. `https://httpd.apache.org/docs/2.4/mod/mod_autoindex.html`, Visited last on 06.06.2017

QueryArrow: Semantically Unified Query and Update of Heterogeneous Data Stores

Hao Xu
University of North Carolina at Chapel Hill
xuhao@renci.org

Ben Keller
University of North Carolina at Chapel Hill
kellerb@renci.org

Antoine de Torcy
University of North Carolina at Chapel Hill
adtorcy@renci.org

Jason Coposky
University of North Carolina at Chapel Hill
jasonc@renci.org

ABSTRACT

Modern system applications often need to interact with metadata from multiple, heterogeneous data stores. An ad hoc solution for integration of multiple data stores by issuing individual statements in the languages of the data stores runs the risk of semantic incompatibilities. This paper describes QueryArrow, a generic software that provides a semantically unified query and update interface to multiple types of data stores. QueryArrow has an algebra-based language called QueryArrow Language (QAL), which can be partially translated to languages of different data stores. We describe the design of QueryArrow, the syntax and semantics of QAL, how QAL is translated to languages of different data stores, and demonstrates its applications as an iRODS database plugin.

1. INTRODUCTION

Modern system applications often need to interact with data from multiple, heterogeneous data stores. There are several recurring tasks in such applications, including the aggregation, access control, discovery, and migration of metadata. A software solution for this challenge is tricky because of the diverse range of types of data stores that it must interact with, including, for example, relational databases, graph databases, and document-oriented databases. Different types of data stores have different types of query languages and data manipulation languages, different semantics of the languages, and different levels of capabilities (such as support for features such as regular expressions).

Some existing solutions aim to bridge the gap between these data store, by either creating a unified query language or API, such as SQL++ [9] and UnQL [6], or refitting a current popular query language or API designed for one type of database into other types of databases, such as Presto [1] and Spark SQL [2]. However, there are at least one of the following two drawbacks in the current solutions:

- The solution does not have a formal definition of the semantics of their query language. Therefore, an ad hoc solution, where results are aggregated from multiple data stores by issuing an individual query in the query language of each database, runs the risk of semantic incompatibilities.

- The solution is query-only and lacks bidirectional support for both query and update. Without support to update, we cannot provide an abstraction of data stores that makes data access transparent to client applications where the metadata is mutable.

iRODS UGM 2017 June 13-15, 2017, Utrecht, Netherlands
Copyright ©2017 Hao Xu

Figure 1: Architecture Diagram

A more principled approach would require complete semantic decoupling of client API from underlying data stores where data are allowed to be mutable. This allows migrating the underlying data stores without modification to the client application. This requires a unified query and data manipulation language. Such a language inevitably is a superset of capabilities of actual data store languages. Therefore, we need to integrate the notion of partially supported features and schemata into this framework, so that we can combine multiple data stores with different capabilities into a unified data store which provides full support to the features and schemata required by the client application.

QueryArrow [14, 16] is built on theoretical developments that allow us to create software that is based on rigorous treatment of semantic foundations of a unified query and data manipulation language. Categorical models of databases allow us to represent the same information in both relational databases and graph databases and transform between the two different representations [13], which is used in QueryArrow for automatic generation of graph database translators. QueryArrow Language (QAL) is structured as near-semiring [8], which allows us to model both query and update, and provide general laws for optimization. QueryArrow is similar to Transaction Logic [5], but we give QAL a monadic semantics [4], which is formalized in Coq [7].

This paper introduces the design of QueryArrow, the syntax and semantics of QAL, how QAL is translated to different database languages, and demonstrates its applications.

2. DESIGN OF QUERYARROW

QueryArrow is made up of three elements: the QueryArrow Service, the QueryArrow Language, and the QueryArrow Plugins (QAP), as shown in Figure 1.

- QueryArrow Service: Register QAP and support execution of QAL
- QueryArrow Language: Provide a semantically unified configuration language, query language, and data manipulation language.
- QueryArrow Plugins: Provide mappings between QAL and external data stores.

A QueryArrow instance includes a QueryArrow Service and a composition of QAPs.

There are three types of QAPs. A data store QAP interfaces with a data store by translating QAL to the language of the data store and interpreting the results returned by the data store. A meta QAP allows users to add aggregation, policy and distributed support to our architecture without additional complexity to the core codebase. An in-memory QAP provides various in-memory functionalities. Currently supported QAPs are shown in the Table 1.

A translation QAP enables translation of the QAL back into QAL, according to user defined rewriting rules written in the configuration fragment of QAL. This allows user to define policy on their metadata. This enables the definition of policies such as metadata access control, distribution, and retrieval optimization. A typical QAP composition is show in Figure 2.

QueryArrow can be run in a distributed environment. A remote QAP is implemented to allow using a QAS as a data store.

3. SYNTAX AND SEMANTICS OF QAL
3.1 Syntax

Name	Description
Sum QAP	aggregation
Translation QAP	policy support
Cache QAP	caching
QAS QAP	remoting
Mutable Map QAP	in-memory mutable map
Immutable Map QAP	in-memory immutable map
ElasticSearch QAP	interfacing with ElasticSearch
Neo4j QAP	interfacing with Neo4j
PostgreSQL QAP	interfacing with Postgres
SQLite3 QAP	interfacing with SQLite3
CockroachDB QAP	interfacing with CockroachDB
FileSystem QAP	file system

Table 1: Available QAPs

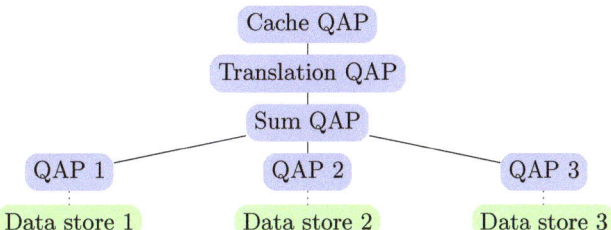

Figure 2: Example QAP Composition

$$
\begin{array}{rcll}
p & & literal \\
prty & & primitive\ type \\
v & & variable \\
P & & primitive\ predicate\ name \\
N & & namespace \\
QP & ::= & P \mid N.QP & predicate\ name \\
ty & ::= & prty \mid \texttt{list}\ ty & types \\
t & ::= & p \mid v \mid [t_1,\ldots,t_n] \mid ty\ t & terms \\
a & ::= & QP(t_1,\ldots,t_n) & atom \\
c & ::= & a \mid \texttt{insert}\ a \mid \texttt{delete}\ a \mid g\ c \mid b \\
& & \mid \mathbf{1} \mid \mathbf{0} \mid c \oplus c \mid c \otimes c & command \\
g & ::= & \neg \mid \exists \mid \texttt{distinct} \mid \texttt{limit}\ n \mid \texttt{order by}\ (\texttt{asc} \mid \texttt{desc}) \\
& & \mid \texttt{let}\ v_1 = s_1,\ldots,v_n = s_n\ (\texttt{group by}\ v_1\ldots v_n)? & aggregation \\
s & ::= & \texttt{max}\ v \mid \texttt{min}\ v \mid \texttt{average}\ v \mid \texttt{sum}\ v \mid \texttt{count} \mid \texttt{count distinct}\ v \mid \texttt{some}\ v & summary \\
\hline
pt & ::= & \texttt{input?}\ \texttt{output?}\ \texttt{key?}\ ty \\
R & ::= & \texttt{rewrite}\ (a \mid \texttt{insert}\ a \mid \texttt{delete}\ a)\ c \\
& & \mid \texttt{predicate}\ P(pt_1,\ldots,pt_n) \\
& & \mid \texttt{import qualified?}\ (\texttt{all} \mid P_1,\ldots,P_n \mid \texttt{all except}\ P_1,\ldots,P_n)\ \texttt{from}\ N \\
& & \mid \texttt{export}\ P_1,\ldots,P_n \\
& & \mid \texttt{export qualified?}\ (\texttt{all} \mid P_1,\ldots,P_n \mid \texttt{all except}\ P_1,\ldots,P_n)\ \texttt{from}\ N & configuration \\
prog & ::= & R_1 \ldots R_n & program
\end{array}
$$

Figure 3: QAL Syntax

The syntax is given in Figure 3. A term t is either a literal p, a variable v, a list of terms, or a type coercion. An atom a is $QP(t_1,\ldots,t_n)$ where QP is a predicate name. c includes four primitive commands: a query command a, an insert command $insert\ a$, a delete command $delete\ a$, and an aggregation command $g\ a$, where g is functions such as max, min, average, sum, count, and count distinct, limit results to first n results, order results by ascending order, order results by descending order, return distinct results, test that results does not exist, test that results exists, or keep certain columns in the results. A command c is either a primitive command or one of the four composite commands: skip $\mathbf{1}$, stop $\mathbf{0}$, choice $c \oplus c$, sequencing $c \otimes c$.

In addition to commands, QAL also allows declaring new predicates and specifying rewriting rules, which are essential for defining policies. The rewriting rules allows us to rewrite a query command, an insert command, or a delete command to arbitrary commands. Also, we have import and export statements. The details of these declarations are given in [16].

Examples applications of QAL are described in Section 5.

3.2 Semantics

The abstract semantics of QAL is formally specified in [15] using Coq. In this subsection, we give a brief description of the formalization.

The key to the formalization is specifying various Haskell typeclasses including `Functor`, `Applicative`, `Monad`, `Traversable`, `Alternative`, `Foldable`, and `Monoid`. We also defined their instances. In addition, we specified `NearSemiRing`. We model all of these in terms of setoids.

The semantics of commands are given in a "store-and-heap" monad, written `sh`, which is a specialization of the `ContT` monad transformer. We define a semantic equivalence relation between commands in terms of this semantics. We have proved that not only is `sh` a monad, but also that QAL forms a near-semiring $(\mathcal{L}(c), \mathbf{0}, \mathbf{1}, \oplus, \otimes)$, where $\mathcal{L}(c)$ is

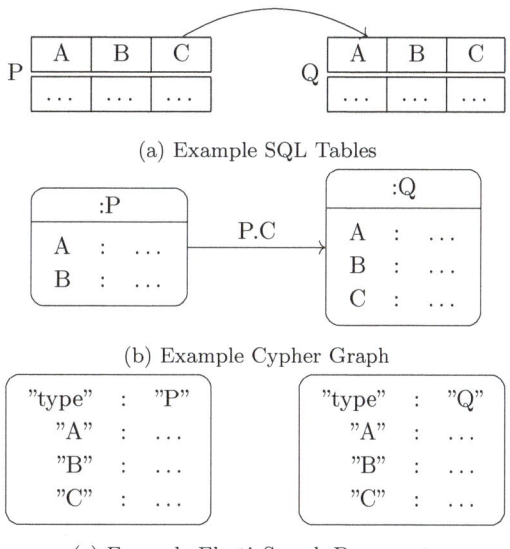

Figure 4: Examples

the language generated by c. This allows us to give an interpretation of **0**, **1**, ⊕, and ⊗. Other language constructs are interpreted in the monadic instance of `sh`.

One challenge to the applicability of this abstract semantics is how to incorporate concrete semantics of different data stores. Inevitably some QAL commands cannot be translated to one statement of one data store. In this case, the QAL command is partially translated, and each subexpression that can be translated into one statement of one data store is executed, and the non-translatable part are executed by a generic execution function. To illustrate this, consider the following example. Suppose that we have three data stores, exporting predicates P, Q, and R, respectively. When a user issues a command such as $P(x, a) \otimes (Q(x, y) \oplus R(x, y))$, the predicates are translated into the languages of the respective data stores, dispatched, and the results are collected and combined according to the abstract semantics. How do we integrate the semantics of different data stores, thereby giving semantics to the whole command? The solution is using built-in commands: we can specify statements in the languages of data stores as built-in commands and plug them into the abstract semantics, as long as they can be interpreted in the `sh` monad. For each language $\mathcal{L}$ of a data stores, a translator from QAL then is a partial function from commands to that language $trans : \mathcal{L}(c) \to \mathcal{L}$. The `sh` monad is designed such that it is parametric to module types `AbstractStore` and `AbstractHeap`. This way, we can choose different concrete implementations of these data structures for different combinations of data stores. An ongoing effort is to integrate SQL into this abstract semantics.

4. TRANSLATION

In this section we list three types of data stores and how QAL is translated into their statements. Our goal is to translate the same commands into multiple data stores which are semantically equivalent based on the concept of observational equivalence: informally, given two data stores, define a relation R of related states of the data stores, if a translation of every successful command from QAL to the languages of the two data stores takes related states to related states, and returns the same set of results, then we say that under the translation, the two data stores are observationally equivalent. For a formal treatment of bisimulation, see [11]. In this section, we assume x, y, z are variables, and a, b, c are primitive values.

4.1 Relational Database

SQL is the query language and data manipulation language for relational databases. In the PostgreSQL QAP, a primitive command is translated to an SQL statement. Users are allowed to defined arbitrary predicates and translations. In addition, QueryArrow provides an automatic translation generator based on a database schema to reduce the coding needed to create QAPs. We illustrate the translation in an example as show in Figure 4a, where $P.A$ and $Q.A$ are primaries keys and $P.C$ is a foreign key to $Q.A$. The automatic translator generates six predicates in two categories: object predicates $PA(x)$ and $QA(x)$, and property predicates $PB(x,y)$, $PC(x,y)$, $QB(x,y)$, and $QC(x,y)$. Object predicates and property predicates differ in how they are translated in insert and delete commands. Object predicates are translated to INSERT and DELETE statement. Property predicates are translated to UPDATE statements. For example, we translate PB as follows:

- query $PB(a,x)$: SELECT B FROM P WHERE A = a
- insert $insert\ PB(a,b)$: UPDATE P SET B = b WHERE A = a
- delete $delete\ PB(a,b)$: UPDATE P SET B = NULL WHERE A = a

and QA as follows:

- query $QA(x)$: SELECT A FROM Q
- insert $insert\ QA(a)$: INSERT INTO Q (A) VALUES (a)
- delete $delete\ QA(a)$: DELETE FROM Q WHERE A = a

Composite commands are translated as nullary and binary partial functions that combine SQL queries. This allows us to, for example, insert a table with columns with NOT NULL constraints. For example, if $P.B$ is not null, then the translation of $insert\ PA(a)$,

INSERT INTO P (A) VALUES (a)

is not valid SQL insert. And $insert\ PB(a,b)$ is translated to

UPDATE P SET B = b WHERE A = a

But $insert\ PA(a) \otimes insert\ PB(a,b)$ should be translated to

INSERT INTO P (A,B) VALUES (a,b)

which is valid.

4.2 Graph Database

Cypher is the query language and data manipulation language for Neo4j [3]. In the Neo4j QAP, a primitive command is translated to an Cypher statement. A graph schema is automatically generated by a translation generator from a SQL schema by QueryArrow as follows, following a variation of the mapping given in [12]: Each table is translated into a node and each column is translated into one of the three: A primary key is translated into a property of the node. A foreign key is translated into an edge. Other columns are translated into a property. Special clauses are added to the translation of insert commands to ensure that primary keys are unique. For example, given SQL table schema as shown in Figure 4a, we can generate a model as shown in Figure 4b.

We may translate PB as follows:

- query $PB(a,x)$: MATCH (n:P) WHERE n.A = a RETURN n.B
- insert $insert\ PB(a,b)$: MATCH (n:P) WHERE n.A = a SET n.B = b

- delete $delete\ PB(a,b)$: `MATCH (n:P) WHERE n.A = a SET n.B = NULL`

and QA as follows:

- query $QA(x)$: `MATCH (n:Q) RETURN n.A`
- insert $insert\ QA(a)$: `MERGE (n:Q{A:a})`
- delete $delete\ QA(a)$: `MATCH (n:Q) WHERE n.A = a DELETE n`

One subtle issue is failure modes. The semantics of relational database and graph database usually do not match. For example, in some graph databases, users are not able to specified unique properties or required properties. Therefore, a translation of $insert\ PA(a)$ is a valid statement, even if the key value a already exists. There are currently two solutions. We can simulate. For example, in the unique property case, we can simulate using the `ON CREATE` clause. We can also create an abstraction layer in which both types of databases implement the same semantics. For example, in the required property case, we can create a predicate $P(x,y,z)$ and hide the more primitive $PA(x)$. This can be done using rewriting rules, import, and export features of the QAL.

4.3 Document-oriented Database

ElasticSearch provides a JSON-based query language and data manipulation language which are radically different from traditional databases. In particular, ElasticSearch is an example of Document-oriented Database. In this type of databases, each predicate is naturally translated to a partial document. For example, given SQL table schema as show in Figure 4a, we can generate a model as show in Figure 4c. This model, unlike the Cypher's case, is often weaker than the SQL model, because it doesn't explicitly encode relations. The translation of predicates is similar to that of graph databases. Because of lack of operators like SQL's `JOIN` or `UNION`, the translation function is mostly undefined on commands with nontrivial combinations of composite commands.

5. APPLICATION EXAMPLES

iRODS QueryArrow database plugin enables iRODS [10] to use QueryArrow as iCAT. This enables the following application examples.

Metadata Access Control. iRODS allows users to tag data object with metadata in the forms (attribute, value, unit) triples. The data management solution stores such metadata in a relational database and is not designed with metadata access control. QueryArrow allows us to add metadata access control using QueryArrow, by defining rewriting rules, without changing the schema of the original database. The extra information such as access control list (ACL) is stored an external database and integrated into the application by QueryArrow.

Metadata Migration. In iRODS, metadata are stored in a relational database. QueryArrow allows us to migrate part of the metadata into a graph database.

Metadata Indexing. As the number of data objects grows, regular queries for data objects become slow. QueryArrow allows us to write rewriting rules so that some of the metadata are put into a search engine based on their attribute name. When user add or remove a metadata item with an indexed attribute name, it is added or removed to the search engine. When the user queries data object by those metadata attribute names, the search engine is utilized accelerate the query.

All of this is transparent to the client application.

6. SUMMARY

In this paper, we introduced design of QueryArrow, the syntax and semantics of QAL, how QAL is translated, and its application examples in iRODS.

REFERENCES

[1] `https://prestodb.io/`.
[2] `http://spark.apache.org/sql/`.
[3] `https://neo4j.com/`.
[4] *Computational Lambda-Calculus and Monads*, 1989.
[5] Anthony J. Bonner and Michael Kifer. Transaction logic programming. In *International Conference on Logic Programming*, pages 257–279, 1993.
[6] Peter Buneman, Mary F. Fernandez, and Dan Suciu. UnQL: a query language and algebra for semistructured data based on structural recursion. *VLDB Journal: Very Large Data Bases*, 9(1):76–110, ???? 2000.
[7] Adam Chlipala. *Certified Programming with Dependent Types : A Pragmatic Introduction to the Coq Proof Assistant*. 2013.
[8] Jules Desharnais and Georg Struth. Domain axioms for a family of Near-Semirings. In José Meseguer and Grigore Roşu, editors, *Algebraic Methodology and Software Technology*, volume 5140 of *Lecture Notes in Computer Science*, pages 330–345. Springer Berlin Heidelberg, 2008.
[9] Kian W. Ong, Yannis Papakonstantinou, and Romain Vernoux. The SQL++ query language: Configurable, unifying and semi-structured, December 2015.
[10] Arcot Rajasekar, Reagan Moore, Chien-Yi Hou, Christopher A. Lee, Richard Marciano, Antoine de Torcy, Michael Wan, Wayne Schroeder, Sheau-Yen Chen, Lucas Gilbert, Paul Tooby, and Bing Zhu. iRODS primer: Integrated rule-oriented data system. *Synthesis Lectures on Information Concepts, Retrieval, and Services*, 2(1):1–143, January 2010.
[11] J. Rutten. Universal coalgebra: a theory of systems. *Theoretical Computer Science*, 249(1):3–80, October 2000.
[12] David I. Spivak. Simplicial databases. April 2009.
[13] David I. Spivak. Functorial data migration. *Information and Computation*, May 2012.
[14] Hao Xu. `https://github.com/xu-hao/QueryArrow`, 2017.
[15] Hao Xu. `https://github.com/xu-hao/CertifiedQueryArrow`, 2017.
[16] Hao Xu, Ben Keller, Antoine de Torcy, and Jason Coposky. Queryarrow: Bidirectional integration of multiple metadata sources. *8th iRODS User Group Meeting, University of North Carolina at Chapel Hill*, June 2016.

www.ingramcontent.com/pod-product-compliance
Lightning Source LLC
Chambersburg PA
CBHW051157220526
45473CB00003B/808